Philosophical Studies Series

Volume 150

Philosophical Studies Series aims to provide a forum for the best current research in contemporary philosophy broadly conceived, its methodologies, and applications. Since Wilfrid Sellars and Keith Lehrer founded the series in 1974, the book series has welcomed a wide variety of different approaches, and every effort is made to maintain this pluralism, not for its own sake, but in order to represent the many fruitful and illuminating ways of addressing philosophical questions and investigating related applications and disciplines.

The book series is interested in classical topics of all branches of philosophy including, but not limited to:

- Ethics
- Epistemology
- Logic
- Philosophy of language
- Philosophy of logic
- Philosophy of mind
- Philosophy of religion
- Philosophy of science

Special attention is paid to studies that focus on:

- the interplay of empirical and philosophical viewpoints
- the implications and consequences of conceptual phenomena for research as well as for society
- philosophies of specific sciences, such as philosophy of biology, philosophy of chemistry, philosophy of computer science, philosophy of information, philosophy of neuroscience, philosophy of physics, or philosophy of technology; and
- contributions to the formal (logical, set-theoretical, mathematical, information-theoretical, decision-theoretical, etc.) methodology of sciences.

Likewise, the applications of conceptual and methodological investigations to applied sciences as well as social and technological phenomena are strongly encouraged.

Philosophical Studies Series welcomes historically informed research, but privileges philosophical theories and the discussion of contemporary issues rather than purely scholarly investigations into the history of ideas or authors. Besides monographs, *Philosophical Studies Series* publishes thematically unified anthologies, selected papers from relevant conferences, and edited volumes with a well-defined topical focus inside the aim and scope of the book series. The contributions in the volumes are expected to be focused and structurally organized in accordance with the central theme(s), and are tied together by an editorial introduction. Volumes are completed by extensive bibliographies.

The series discourages the submission of manuscripts that contain reprints of previous published material and/or manuscripts that are below 160 pages/88,000 words.

For inquiries and submission of proposals authors can contact the editor-in-chief Mariarosaria Taddeo via: mariarosaria.taddeo@oii.ox.ac.uk

Manhal Hamdo

Epistemic Thought Experiments and Intuitions

Springer

Manhal Hamdo
Institut für Philosophie
Freie Universität Berlin
Berlin, Germany

ISSN 0921-8599 ISSN 2542-8349 (electronic)
Philosophical Studies Series
ISBN 978-3-031-33479-5 ISBN 978-3-031-33480-1 (eBook)
https://doi.org/10.1007/978-3-031-33480-1

This Springer imprint is published by the registered company Springer Nature Switzerland AG
The registered company address is: Gewerbestrasse 11, 6330 Cham, Switzerland

Introduction

This work is a defense of the nature of intuition as a mental state and its use in epistemology. The everyday usage of intuition-terminology applies to more than such states. For instance, a parent might have an intuition that their child is highly intelligent, or a student might have an intuition that certain parts of the syllabus are important for exams. The focus of this book is different: it is the study of intuition which serves an argumentative role in epistemic thought experiments, as in Jackson's intuition that our physical knowledge about Fred does not help us to know everything about his color experience, that Mary's previous physical knowledge is incomplete, and Searle's intuition that any computer lacks cognitive states.

Recently, much has been written about the topic of intuition, its nature, reliability, or epistemic status. There are common disputes on intuition both on whether or not we should appeal to intuition in doing philosophy as well as on the question of what 'intuition' means. These debates have persisted for many years. Different philosophers have different attitudes toward the nature of epistemic intuition. Similarly, there is no generally accepted theory of the reliability or evidential status of intuition. One at least can differentiate between three main positions on the epistemic status of intuitions—the reliablists, skeptics, and perspectival relativists.

This book attempts to give an account of what intuition is and what is its epistemic status. In doing so, it will locate these views in the current landscape of discussion. I will argue that my account of intuition has advantages over other theories of intuition and its epistemic status. In the book, I will be developing an account of epistemic intuitions which correctly captures the nature of the intuition used in epistemology and demonstrates how epistemologists can best use those intuitions to justify their epistemological theorizing.

This book is divided up into four somewhat lengthy chapters, excluding the introduction and conclusion.

In the first chapter, I put the subject matter into perspective. To that end, I will look at the progenitor of epistemic intuitions, namely, thought experiments. I will initially be discussing how, generally speaking, thought experiments play a crucial role in different fields, including philosophy. Later, I shall go on to specifically focus on epistemic thought experiments discussing some paradigmatic examples of thought experiments. As will become clear, an epistemic thought experiment is an imaginative, intentional, and concrete or possible scenario that takes a narrative form of a postulated development of events and can be best analyzed as an epistemic argument accompanied by a crucial epistemic significance of informativeness relative to the thought experimenter's epistemic state. And such an argument is directed at criticizing, establishing, or simultaneously criticizing and establishing a theory. It does so mainly in virtue of the help offered by two unique components: an idealized assumption and the intuition arising from it, producing either justifiedness or unjustifiedness of the pertinent theory.

In light of the above, the thought experiment detects the presence of intuition in contemporary epistemology. What I first plan to explore is how intuition terminologically and epistemically has been developed across epistemic discourses from Plato to the present. Consequently, I will be engaging in an attempt to describe and discuss what most intuition-theorists commonly identify as intuition. I shall be arguing that their mathematical conception of intuition amounts to Platonism about intuition. I shall argue that Wittgenstein's anti-Platonic arguments in the philosophy of mathematics may help to successfully avoid such Platonism, but they leave us with an epistemology without intuition as a mental state.

In the second chapter, I will specifically be developing and defending what I see as the correct conception of epistemic intuitions. I explore and discuss three epistemic treatises given by recent epistemologists who identify intuition as a mental state. Those accounts are Bealer's intellectual seeming analysis, Sosa's understanding view, and Chudnoff's account of intuition as intellectual perception. The preceding accounts mainly ascribe three features to intuition. (1) Contrary to experimentalists who advocated the thesis that intuition is knowable a posteriori since it can easily be affected by variables irrelevant to the matter the thought experiment seeks to address, apriorists argue that it is knowable in an a priori sense. (2) Intuition is a propositional attitude (i.e., a mental state held by a subject toward a proposition). (3) Intuition is phenomenal (i.e., what it is like to be intuiting). Maintaining that intuition is a mental state, I address each of those three features in turn.

First, I plan to show that the state of intuition is best understood as enshrined in a multi-dimensional space of appraisal. To that end, I first introduce two possible arguments against apriorism about intuition. One opposes Chudnoff's argument of the number sequence, according to which intuition is a priori because it is part of our abstract knowledge that is purified or free from any sensory or experiential elements. However, an opponent of Chudnoff's line of argument may argue that the number sequence may be just a result of memory, testimony, instruction, induction, or counting technique we learned by practice. If one or all of the aforementioned probabilities is true, then experience in one way or the other is involved in our knowledge of the number sequence. Therefore, Chudnoff might be wrong. Another is a response to the common argument from concepts given by Bealer, Sosa, and Chudnoff. According to it, intuition is a priori because it consists of concepts that involve abstract (non-experiential) thought. The problem with this argument is that it is not in sync with recent discoveries by cognitive scientists and cognitive neuroscientists who identify certain perceptual mechanisms that underlie conceptual processing. If so, if intuition consists of concepts that include perceptual construals, then intuition is a posteriori cognition. Yet, as used by philosophers, intuition is to show whether some cases are instances of the concept in question. In this sense, intuition is a priori. Thus, intuition can have both these features. And since there are some borderline cases, as in introspection, which is neither a priori nor a posteriori, if an intuition emerges from them, it will be introspective intuition and accordingly neither empirical nor rational. Therefore, I have come to the conclusion that intuition might have the aspect of being a priori, a posteriori, both a priori and a posteriori, or neither a priori nor a posteriori. That is to say, intuition is multi-dimensional and cannot be restricted to a certain way of knowing.

Second, I move on to argue that the state of intuition is intentional though non-propositional. A proposition is the primary meaning of a sentence referring to one or all of the following: the bearer of truth and falsity and nothing else, the referent of a that-clause, or the object of a mental state. Such states are called propositional attitudes. By a propositional attitude, it was meant that a mental state articulates a relation between the subject and proposition. Bealer, Sosa, and Chudnoff adopt this view and regard the state of intuition as propositional.

Contrary to that, I show that propositions do not always take the form of a that-clause. Nor do characterizing mental states as propositional attitudes solve the long-term puzzle of empty names. For mental state, verbs can be about a name, an object, a number, or a proposition without predicating anything about them. Being directed to an object, mental state verbs may be intentional or objectual but not propositional. However, for a mental state to be propositional, it is not enough to be objectual. Rather, it must have propositional content, i.e., to be sensitive to the truth-conditions. And this is not available due to reasons given by Cohen and Goldman. According to Cohen, the truth-conditions vary according to the inductive ground. And, for Goldman, they vary according to the distinguishability or discriminability of the situation. Therefore, intuition is not propositional.

Next, I will show that intuition has no special phenomenology. My strategy to achieve this aim lies in the argument from the case studies, according to which thought experiments considered in the first chapter provide evidence that no mention of a distinctive phenomenology is required to explain the emergence of the relevant intuitions. At no point in these thought experiments does the special phenomenology of intuition play any role. The authors are unaware of such special phenomenology, and so they certainly did not mention either explicitly or implicitly that such special feeling accompanied their intuitions. As a result, the presence of a special phenomenology that allegedly accompanies intuition is not present in the original texts where cases were first presented.

Lastly, I conclude that intuition is not individuated in terms of phenomenal character, propositional attitude, and a priori-based view of conceptual competence. Instead, it is individuated in terms of the point of its origin, thought experiment, and has at least a lot in common with it, such as directedness, imagination, informativeness, being either controversial or universally accepted, and possessing a strong idealization.

Once I have developed and defended my account of the nature of epistemic intuitions, I will move on in the third chapter to show how those intuitions have epistemic weight—that they can be used to produce justified epistemic arguments. To show how intuitions have epistemic or evidential weight, I will critically look at three positions with regard to the question of the epistemic weight of intuitions—the reliablists, skeptics, and perspectival relativists.

Reliabilism concerning the epistemic status of intuitions maintains that intuitions are essential to epistemology and that one's having an intuition that p provides epistemic support for one's being justified in believing that p or for one to know that p. Such a position is mainly maintained by Bealer, Sosa, Chudnoff, and others. While there are slight differences between the various advocates of reliabilism, reliabilism about intuitions remains the main thesis that intuition evidentially qualifies as a reliable source of knowledge. However, in addition to expository prologues to these three views, I will be trying, at least, to undermine Bealer's reliablist form of modal intuition, Sosa's competence-based view, and Chudnoff's phenomenal dogmatism.

To that end, I shall raise some difficulties with Bealer's proposal of semantically stable philosophical terms such as intuition, its modality, and classifying the scientific essentialism attack under the category of local scientific essentialism whose concepts are not central to philosophy. Similarly, Sosa's proposal will be shown as problematic since it does not explain by virtue of what the competence he talks about makes epistemic intuitions reliable. Nor does it specify a standard on the basis of which we can verify our intuitions unless there is no dialectical-epistemic dichotomy. Lastly, I will give a demonstration of how Chudnoff's view raises more difficulties than it solves. For it does not explain why such alleged presentational phenomenology, if any, is relevant to the epistemic status of intuition at all. Chudnoff cannot place much weight upon his assumed phenomenological analogy between intuition and perception because even if such analogy holds, phenomenology does not save perception from the possibility of error which is verifiable in perception but not in intuition.

On the other hand, skepticism concerning the epistemic status of intuitions maintains that intuitions are epistemologically useless. There is something either about variation in intuitions or the lack of ability to calibrate our intuitions that leads the skeptics to the conclusion that intuitions have no epistemic weight and should be excluded from philosophy. In this regard, I will assess two arguments in support of skepticism about the epistemic status of intuitions: the variation argument and the calibration argument. Then, I will show that both of these arguments are doubtful. For, despite the fact that skeptics' attack is an attempt to make a transition from a few studies to a very large statement, agreement is a self-defeating criterion by which something may be judged as true or false. And, if intuition does not calibrate, then it follows that skeptics' method cannot be an admitted method to judge whether or not intuition qualifies as a reliable source of evidence in epistemology. It also entails that we cannot know which intuition is reliable and what is not.

Lastly, perspectival relativism holds that intuitions ought to and do count as evidence in support of some claim, but we need to rethink how this happens in light of considerations derived from three perspectives or methods in the process of wide reflective equilibrium; especially that none of these methods is superior over others and accordingly their truths have the same epistemic status. But I render this position questionable. For worries crop up when it is said that Christian revelation and ritual use of hallucinogens are philosophical methods. If so, if they are not, we are left with intuition as a single method on which philosophy relies. But this is imprecise. Philosophers often rely on argument in their theorizing, for instance. However, subjecting intuition to reflective equilibrium does not necessarily lead to truth; it may lead to epistemic foundationalism or conservativism.

Having undermined the main positions on the epistemic status of intuitions, I will try to articulate what seems to me to be the right position that any adequate account must respect. The suggestion is, in fact, unlike the above analyses of the epistemic status of intuition, that intuition alone has no epistemic status. Yet, it suggests that intuition has evidential weight. That is to say, intuition combined with other pieces makes up a whole. One part of the whole is the thought experiment that generated intuition. If so, the question of epistemic status does apply to the whole and not the intuition alone. I will attempt to show that such a whole constructs an epistemic argument. What is more, I shall argue that epistemic argument is different from the logical argument, and thereby they must be appraised differently both in terms of the standards of appraising as well as the appraisal words.

So conceived, this view is generally resistant to the errors of previous types of views. For it maintains the epistemic status of intuition without considering intuition to have epistemic weight in virtue of its being phenomenal, propositional, or a priori. The suggestion is that intuition can have epistemic status and be calibrated by virtue of its ability to fit in well with other pieces to construct a justified or good argument that is understood in epistemological terms. If so, then whether intuition is semantically stable, modal, competent, dialectically significant, can be subjected to reflective equilibrium, or so on seems to be irrelevant to its evidentiality. Nor does the epistemic status of intuition analyzable in terms of the intuiter's conceptual competence.

In the last chapter, I assess the quality of the account of the nature and epistemic role of intuition defended in the previous chapters in light of some of the main current objections to the centrality of intuitions. Since the literature on this topic is huge and has lately been increasing at an exponential rate, I will only attempt to look at a few key objections. Specifically, I will focus on the following views: the thesis of anti-centrality given by Herman Cappelen, the philosophical method of the counterfactual conditional given by Timothy Williamson, and Max Deutsch's view of anti-intuition-based methods of analytic philosophy.

While there are crucial differences between Williamson, Deutsch, and Cappelen, they all are inclined to adopt the extreme view that intuition never plays any evidential role in philosophy. More precisely, they mainly argue that (1) due to the disagreement among philosophers at the level of the concept of intuition, they express skepticism toward whether there is a referent of the term 'intuition' at all; they also argue (2) against psychologizing the evidence in philosophy: intuition is a psychological event, and by considering it as evidence in philosophy, one psychologizes the idea of philosophical evidence; and (3) they attempt to prove that philosophers never boost their views by means of intuition but always via argument.

I shall be responding to these three major intuition-rejecting arguments in these accounts: the argument from lacking reference, the argument against psychologizing the philosophical evidence, and the argument from *argumentationalism*. However, I will show that they fail to affect the account of intuition I attempted to establish in the previous chapters. If I do so successfully, then those intuition-deniers are wrong, and thereby my account is secured.

Concerning the argument from lacking reference, Cappelen argues that philosophers have never shown a pattern of agreement on what intuition is, and accordingly, they never rely on it as evidence. To this, one can respond that philosophers can rearrange their claims in terms of some less controversial category of intuition-terminology, like System 1 reasoning to mean what is intuitive and System 2 reasoning to mean what is not. In addition, his argument is too theory-laden. He assumes an overly thick theory of intuition. That is, intuition has a distinctive phenomenology, being based only on conceptual competence, and offering fundamental justification. While my concept of intuition falls under the thin category of intuition and not the thick one. It does not require intuition to have any of the above-mentioned characteristics.

And unlike Williamson, who notes that philosophers vary in labeling intuition either as a priori, or empirical, or phenomenal, or non-inferential, I will argue, in my book, that intuition is multi-dimensional, non-phenomenal, and inferential. That being so, my account is not what he planned to deny.

According to Deutsch, treating intuitions as evidence can be understood either in the sense of CONTENT intuitions (relying on intuition, without considering it as evidence that it is true) or in the sense of STATE intuitions (relying on intuition as evidence that is true). He states that it is a myth to claim that philosophers rely on the sense of STATE intuitions as evidence. It is true that, in the picture given in my book, the focus will be on the sense of STATE intuitions. But I will also be arguing that in order for a state like intuition to be considered as evidence, one must have such intuition, and additional conditions (e.g., being combined with other pieces to make up a whole) must be added. That being so, it is misleading to allegedly extend Deutsch's anti-centrality theory by making it applicable to my particular sense of intuition as a state.

Regarding the argument against psychologizing the philosophical evidence, anti-centralists suppose a propositional theory of philosophical evidence that consists of non-psychological propositions and facts such as beliefs, inclination to believe, etc., about intuition. Despite the fact that I am not going to suppose a propositional view of intuition, my account will imply that the evidential status of intuition does not depend on intuition itself; rather, it rests on its being combined with other pieces to make up a whole—thought experiment or argument. Hence, even if I might not think of intuition as ultimate evidence, I typically agree that intuitions can at least be derivative evidence. And the assumption of centrality does not require that the evidential role of intuition should be ultimate. Such a version of centrality is not one that supporters of the anti-centrality plan to reject.

Concerning the argument from *argumentationalism, intuition-deniers* analyze a series of canonical epistemic thought experiments and conclude that philosophers use argument rather than intuition as evidence in their evaluation of philosophical theories. For instance, in his analysis of Chalmers' Zombie Case, Cappelen gives special importance to the absence of the aforementioned typical features of intuition and the presence of argument instead. Avoiding repetition at the level of my view concerning the typical features of intuition, my focus here will be mainly on the presence of argument, proposing an amendment, according to which intuition and argument are not mutually opposed, inconsistent, or exclusive. Intuition is part of the structure of epistemically justified argument and figures in it.

What all this entails is that my version of the nature of intuition and its justificatory role remains untouched by anti-centrality objections.

Acknowledgments

This work has been a long time coming, and my greatest debt is to those who have helped me to think through the material in it. I have received enormous encouragement and support from so many for so long—in conversation, through email, over dinners, at talks, etc. I would like to thank, first and foremost, Dr. Nilanjan Bhowmick for his never-ending input and advice throughout the entirety of my journey of writing this book. Without his invaluable support, belief in me, patience, thorough, careful, insightful criticism, constant guidance, and encouragement, this book would not have come to fruition. However, any improvements in this work are, I would like to emphasize, attributed to him; any errors are solely mine. Dr. Bhowmick has not only helped me through this project but also has shown me what it means to be a researcher and a good learner. I can never adequately express my gratitude.

Many big thanks to all the professors, teachers, colleagues, and friends who listened to my interminable talks about intuitions, who put up with my sometimes ridiculous ideas, who argued with me when I needed arguing, and who helped keep my head above water. Many, many thanks to the friends whom I did not talk philosophy with and who consequently helped keep me sane.

My thanks go to two anonymous reviewers of the manuscript for Springer Nature, who made many penetrating comments on the same. To my editor at Springer Nature, Christopher T. Coughlin: Thanks for keeping the faith.

Finally, I would like to thank my parents, whose unconditional love, faith, and support have been a constant treasure, not just through the writing of this work but throughout my entire life. They have been the center and the ones to give me a hand when I needed it. So, I love you, and thank you. To Shiv Kumar, I owe a great debt of gratitude. I am grateful for his endless kindness, forgiveness, and unwavering moral support during my most discouraging and frustrating hours. I am indescribably lucky to have him by my side.

Manhal Hamdo

Acknowledgements

[illegible]

Abstract

This work is an attempt to investigate the nature of intuitions demonstrating how philosophers can best use them in epistemology. First, I consider a number of paradigmatic thought experiments in epistemology that depict the appeal to intuition and demonstrate that the nature of thought experiment-generated intuitions is not best explained by an a priori Platonism. Second, I develop and argue for a thin conception of what epistemic intuitions are. This account maintains that intuition is neither a priori nor a posteriori but multi-dimensional, an intentional but non-propositional mental state, non-conceptual and non-phenomenal in nature, which is individuated in virtue of its progenitor, namely, thought experiment. Third, I provide an argument for the evidential status of intuitions based on the correct account of the nature of epistemic intuition. The suggestion is the fitting-ness approach: intuition alone has no epistemic status. Intuition has evidentiary value as long as it fits well with other pieces into a whole, namely, thought experiments. Finally, I address the key challenges raised by supporters of anti-centrality, according to which philosophers do not regard intuition as central evidence in philosophy. To that end, I respond to them, showing that they fail to affect the account of intuition developed in this book.

Contents

Chapter 1
Epistemic Thought Experiment and Intuition

1.1 Introduction

This first chapter aims to start putting the subject matter, the nature and epistemic status of intuitions, into perspective. Intuition forms a central aspect of contemporary theories in epistemology. Thought experiments are often used to elicit intuitions and arrive at judgments that support philosophical theories of one hue or the other. However, the problem is with establishing the claims made using intuitions. Epistemologists just take them for granted without offering a characterization of what it is to intuit. The nature of intuitions is not clear, and neither is it beyond doubt. I will articulate the nature of intuitions by investigating a number of paradigmatic thought experiments from various written texts of contemporary epistemology that would be indicative of an appeal to intuition.

That being the case, before launching into intuitions themselves, I will take a look at their generator, thought experiments. To that end, I will be showing how, generally speaking, thought experiments play a crucial role in different fields, including philosophy. Subsequently, I shall move on to focus on epistemic thought experiment discussing some representative examples of thought experiment from classical epistemology. Doing this will show that even though epistemic intuition is still a relatively young topic in epistemology, it is not a new phenomenon. Its roots in epistemology are deep enough. It is important to demonstrate this fact as it will set the stage for the next step, namely, discussing in more detail some paradigmatic cases from contemporary epistemology. The direct result of doing all this is that it will allow us to depict what epistemic thought experiments are like and what is their structure and function.

In fact, however, this understanding of the appeal to intuition as present in recent epistemic thought experiments has a twofold impact: an empirically founded view of the source of epistemic intuitions in general and paving the road to the two important questions which will be addressed subsequently in this and the next chapters of

M. Hamdo, *Epistemic Thought Experiments and Intuitions*, Philosophical Studies Series 150, https://doi.org/10.1007/978-3-031-33480-1_1

this book. One question concerns the nature of epistemic intuition. Another question concerns the epistemic role of intuition. In order to answer the first question, I will start attempting to explore how intuition terminologically and epistemically has been developed across epistemic discourses from Plato to present.

Nevertheless, exploring thought experiment-generated intuitions in the present literature alone is not enough to pave the way for establishing my own inclusive picture of the nature of intuition, simply because most of the intuition-theorists have different identifications of intuition. That is why I will be engaging in an attempt to describe and discuss what most of the theorists about intuition in the current literature identify as intuition. This discussion will show that thought experiment-generated intuitions with which I am concerned here, and intuition-theorists' intuitions are different. I will argue that the latter brings about a Platonic view of intuition, which in turn entails that, contrary to what those theorists themselves argue, intuition is not a mental state. I shall also argue that Wittgenstein's anti-platonic arguments may successfully avoid such Platonism but can lead to a picture in which intuition is not a mental state. Thus, I have to start drawing my own portrait of what epistemic intuition is and/or is not. Here is the plan for the chapter.

The chapter is divided into eight parts. In the second part, I will be showing how the usage of thought experiment is diversified among a great number of fields. As well, I will be providing a brief overview of the emergence of the concept of thought experiment and the development of its usage in philosophy over time from antiquity to the present. In the third part, I shall be concerned especially with thought experiments that provide a probative force both in classical and contemporary epistemology. In Sect. 1.3.1, I will be discussing three exemplary cases from classical epistemology. In Sect. 1.3.2, I will be considering ten of the most common thought experiments in contemporary analytic epistemology. In the fourth part, thought experiment itself will be the object of study. We will see how intuition inevitably arises from thought experiment. Accordingly, in the fifth part, I shall discuss in brief the history of the etymology and usage of intuition both in classical and contemporary epistemological inquiry. In the sixth part, I will argue that the theories of intuition offered by intuition-theorists amount to Platonism about intuition. In the seventh part, I shall be showing how Platonism as such can be avoidable by Wittgenstein's view of mathematical intuition, which yet leaves us in epistemology without intuition as a mental state. In the last part, I will sketch some concluding remarks of the chapter.

1.2 The Emergence of Thought Experiments in Philosophy

In the literature to date, thought experiment plays the role of a means of reasoning that practitioners use to investigate philosophical principles or ideas. There is no doubt that thought experiment takes up the attention of scholars in a diverse set of fields such as political thought (Miščević 2018), economics (Schabas 2018), theology (Fehige 2018), ethics (Brun 2018), philosophy of religion (Taliaferro & Knuths

2017), physics (Peacock 2018), biology (Schlaepfer & Weber 2018), and mathematics (Starikova & Giaquinto 2018). This list goes on and on.

Here is a brief sample of how a thought experiment may be used in a physics class. Suppose a teacher invites the students to construct a thought experiment in order to see to what extent their knowledge about the Coriolis effect is correct.[1] The scenario can be described as the following. Suppose that two children are playing by propelling a ball with force through the air by a movement of the arm and hand to each other. They are doing so while riding on a merry-go-round platform and sitting on opposite sides. Now, suppose that one of them hurls the ball straight ahead, relative to his position. The question is whether the other will be able to catch it. Pupils who have the right knowledge about the Coriolis effect said no simply because the ball hurled down by one of them straight ahead curves to the side from the standpoint of the one who is located on the merry-go-round. Thus, students who have the wrong knowledge about the Coriolis effect would say yes.

However, the value of thought experiment far outstrips the use any discipline makes of it. It is also common in philosophy. Accordingly, I turn to some understanding of its history in all domains of philosophy as it is practiced.[2] Although, according to (Witt-Hansen 1976), it was Hans Christian Ørsted who coined the term "thought experiment" in philosophy for the first time in 1811, and thought experiment as a subject of study started eagerly exciting philosophers' concerns in the last few decades only, the history of philosophy and thought experiments are mutually dependent. In other words, thought experiments' role "in philosophy is as old as the subject itself" (Rescher 1991, p. 32). This thesis sounds true to philosophical ears. It has at least initial support from intriguing experimental evidence.

This sort of evidence can be easily offered if we begin with ancient philosophy, where we can find that thought experiments served three key goals, namely, supporting, attacking, or suspending particular philosophical claim(s). These three purposes respectively match three thought experiments. The first thought experiment is described by the Pythagorean philosopher Archytas of Tarentum, who has given the image of a spear thrown across the edge of the cosmos in order to support the thesis of infinite space. The second thought experiment is mentioned by Plato in his *Republic,* namely, the ring of Gyges, in which Plato asks us to imagine a shepherd named Gyges, who was given a ring that enables him to become invisible and anonymous. Gyges uses this invisibility and anonymity to seduce a queen, murder her husband, and take over the kingdom. The point of the story is to indicate, ultimately, the fault of the view that establishes happiness on the basis of desire, injustice, and self-interest. Lastly, Sextus Empiricus' arguments whose conclusions suspend judgment on whether a place is divisible or indivisible. To give emphasis to this view, one can discuss many thought experiments, such as the ship of Theseus, which raises the question of whether the identity of an object is equal to its components. In

[1] The Coriolis effect is an apparent deflection of a moving mass or object in a rotating frame.

[2] I mean to imply that philosophers use thought experiments when doing philosophy. Not exclusively, not always, and perhaps not all philosophers, but most of them and quite often.

the story of Dion and Theon, Chrysippus questions the principle of identity, according to which two different things cannot occupy the same place at the same time, and so on (Ierodiakonou 2018).[3]

However, some differentiate between these thought experiments in Greek philosophy and what is called as fictional creations, which appear particularly in Plato's *Republic*. It is maintained that one can identify mainly three fictional styles: myth, simile, and dialogue. These styles can be related to the myth of Gyges, the myth of Er, and the metaphor of the cave (Becker 2018).[4] Moreover, the main function of thought experiments, according to Aristotle, is to compensate for our shortage of epistemic data, which is needed but cannot be obtained in sufficient amounts (Corcilius 2018). That is to say that when philosophers run out of epistemic data, they appeal to thought experiments that compensate for the lack of it.

It is important to note that, contrary to what some may think, thought experiments were used both in Western as well as Eastern philosophical traditions. For example, we find that Gettier-like cases were discussed in Indo-Tibetan epistemology. In fact, a number of highly respected scholars in Indo-Tibetan tradition like (Matilal 1986 and Dreyfus 1997), for example, hold that those traditions contain some cases similar to Gettier's cases, according to which knowledge is not justified true belief. Just as with Gettier cases, in Gettier-like cases, the subject lacks knowledge.[5] In other words, the subject's mental states or events do not successfully achieve knowledge. However, (Stoltz 2007) expresses his disagreement with this view due to two reasons. One is that what Matilal and Dreyfus consider as true belief in Indo-Tibetan tradition is a cognitive event or episode of factive assessment. But, first, Stoltz argues, "factive assessment is held to be a specific type of mental episode, whereas true belief is not." Second,

> the justified true belief analysis of knowledge is quite remote from Indo-Tibetan epistemology, and so finding a parallel to 'justification' within this context is not straightforward.

---

[3] The ship of Theseus is a thought experiment that has been discussed by many philosophers such as Heraclitus, Plato, Plutarch, etc. The paradox's basic strategy is to raise the question of whether an object remains the same after replacing all its parts.

Dion and Theon is a thought experiment that was posed by the Stoic philosopher Chrysippus as follows. Yesterday, there was a whole-limbed man called 'Dion,' who had a part, which consists of all of Dion except his left foot, called 'Theon'. Today, Dion's left foot was cut off. If the two individuals are still alive, they are two different objects residing in the same place and wholly made up of the same matter.

[4] In the Myth of Er, Plato writes about a soldier called Er, who is dead and revived to describe the afterlife where the moral or just people are rewarded and the immoral or wicked people are punished.

In the allegory of the cave, Plato asks us to imagine chained people who can see only the wall of the cave in which they were imprisoned. Behind them, there is a fire that burns. Between them and the fire, there are puppeteers with puppets that throw shadows on the wall of the cave. What the prisoners see is not reality but only its shadows. Shadows represent the objects in the perceptual world. Puppeteers with puppets represent the world of Forms in which perceptual objects participate.

[5] Gettier cases are two thought experiments given by (Gettier 1963) to question the usual concept of knowledge as justified true belief. In them, the characters have justified true beliefs without knowledge. These cases will be discussed in more detail in Sect. 1.3.2.8.

> This problem is compounded by the fact that if we are to attribute theories of justification to Indo-Tibetan epistemologists, many of these theories will need to be externalist in nature. Externalists commonly eschew the justified true belief model of knowledge, however (Stoltz, 2007, p. 396).

At any rate, our concern here is not whether examples from Indo-Tibetan tradition are similar to Gettier's cases or not. What matters for us here is that thought experiments are used in the Indo-Tibetan tradition. Even Stoltz himself agrees with a view as such.

Moving on to medieval Arabic and Islamic philosophy, thought experiments were also employed in the work of Avicenna, Ibn al-Haytham, Abu Hamid al-Ghazali, etc. For example, Avicenna's case of the flying man presents thought experiment as a creative, subtle, internal, sensory faculty that excludes both hypothetical counterfactuals (i.e., *wahm*) as well as Aristotle's hypothetical conditionals that are dependent on imagination and intellect (McGinnis 2018).[6] Avicenna's flying man thought experiment was considered as the ancestral ground of Descartes' cogito. (Oktar 2014) argues that Avicenna, long before Descartes, comes to the same conclusion as Descartes: the existence of a conscious self without the need for the existence of the body.

If we return to the Western tradition, we cannot ignore Galileo, the father of the fundamental turning point in our understanding of the world in the sixteenth century. It is often said that thought experiments had an honored place in Galileo's style of inquiry. However, Palmieri (2018) presents Galileo in an uncommon way: first, there is no division between what is real/material and what is a thought experiment. That is, for him, in Galileo, we can talk about reality through thought experiments. Second, Galileo enjoyed his thought experiments, and did not use them to go into a solution of paradoxes, as in Galileo's case of falling and floating objects (Palmieri 2018).[7] Maybe, this enjoyment is particularly what makes his thought experiment attractive. Thought experiments had a certain aesthetic pleasure for Galileo, apart from their heuristic uses.

Sometimes different philosophers come up with almost the same thought experiment. Typical examples of such thought experiments are those whose subject matter are properties of the physical universe. For instance, Newton's bucket and Leibniz Mill are thought experiments about the nature of matter and motion. Although these thought experiments were also influenced by some ideas given by other philosophers like Descartes, for example, they, in turn, influenced other contemporary philosophers like Searle, who produced the Chinese room case in order to argue against

[6] Flying Man is a thought experiment given by Avicenna to argue for the idea that the existence of an immaterial rational soul or self is totally independent of the existence of the physical body. It asks us to imagine a man suspended in space in such a way that his individual self will lose awareness of his physical body. Yet, it will be absolutely aware of its individual existence.

[7] Unlike Aristotle, Galileo proved through this case study that both heavy and light objects fall at the same speed.

Physicalism (Arthur 2018).[8] Each of these thought experiments branched out to another in a different context and motivation.

Last but not least, let us consider the transcendental thought experiments such as Kant's request for us to think about the absence of space or time, Wittgenstein's many cases in which he asks us to imagine ourselves breaking regularities, etc. These thought experiments play a crucial role in knowing the nature of philosophy and its limits. Westphal (2018) holds that if the subject matter of these thought experiments is what we can/cannot do, then our lacks and obstacles play an essential role in constructing philosophical theory. Recently, the scope of thought experiments has grown considerably to shape the present of philosophy. This can be easily shown with reference to Quine's Gavagai, Davidson's Swampman, Putnam's twin Earth, Thomson's violinist, Foot's trolley problem, and many more.[9] In short, thought experiment is the very bones of philosophy. A question that arises with the connection between philosophy and thought experiment is whether thought experiment plays the same crucial role in epistemology. This is what I will focus on in the next section.

---

[8] A bucket is a thought experiment described by Newton to support his claim that motion can be explained only by absolute space. The experiment describes a bucket of water suspended by a cord that is twisted tightly and then released. At first, the bucket begins to rotate rapidly. Gradually, the water starts spinning even after the bucket rests. Then, the water stops rotating. Eventually, absolute motion accounts for the physical motion of both the bucket and the water. This worldview was debated by his contemporary Leibniz mainly by two arguments. Each argument involves what the so-called Leibniz shift is. Supporting in part Descartes' recognition that motion is both absolute and relative, these shifts hold that space is actually nothing but the extension of bodies. In his Monadology, one can know better about his view about the nature of the body or matter. In there, he introduces what is known as the mill argument, in which he holds that pure material objects like the brain, machine, etc., cannot think, perceive, and so on. This thought experiment is an ancestor to Searle's Chinese room case, according to which, regardless of how well-programmed a computer is, it cannot do what the human mind does. The case will be discussed in more detail in Sect. 1.3.2.3.

[9] Gavagai is an example used by Quine to support his thesis on the indeterminacy of translation. The experiment runs as follows: Quine imagines a native speaker of some unknown language uttering the word "Gavagai" when the native sees a rabbit. What seems natural is that many translations of this word can be given by speakers of English (rabbit, undetached rabbit part, stage of rabbit). This leads to the indeterminacy of translation. Swampman is the protagonist of a thought experiment given by Donald Davidson to support his notion that the causal history of the speaker is what determines the meaning of one's words. Swampman is a creature that is Davidson's duplicate but means nothing by his words due to not having the same causal history. Putnam's Twin Earth thought experiment argues that the meanings of words are not ultimately determined by means of pure psychological terms. Twin Earth is a replica of Earth in everything except that water in it is not H2O but XYZ. Although people on both planets share the same beliefs and images linked with the word water, the term varies in meaning on the two planets; simply because it refers to two different substances. Thomson aims to defend the view that the fetus's right to life is overestimated and that abortion has to be permitted in some cases. To that end, she appeals to a thought experiment in which she analogizes the case of a woman who has no right to abortion to the case of someone who was kidnapped and plugged into someone else to rescue the life of the latter. By this thought experiment, Thomson also aims to respond to Philippa Foot, who questions moral consequentialism, the view that moral action is best explained by its consequences, by introducing the trolley dilemma in which she imagines a driver of a trolley who has no option but either killing a person and saving five people or the other way around. Either way, the consequence is immoral unless the killing is moral.

1.3 Thought Experiments in Epistemology

In the previous section, we stressed the mutual dependence between the history of philosophy (western and eastern) and thought experiments on each other. This usage includes more in the extension of the term 'thought experiment' than we are concerned with here: the use of thought experiment in epistemology. What I want to explore here is thought experiment as "a curios way of gaining knowledge about the world" (Clatterbuck 2013, p. 309). In what follows, I shall discuss such usage that was a common activity at least since Plato's *Theaetetus*, Descartes' case of Evil Demon, and the like.[10] However, thought experiment has been found even more frequently in contemporary epistemology, as in Searle's example of Chinese room, Chalmers's case of Zombie and many more.[11] My principal concern here is to bring out the epistemological use of thought experiment in recent years.

There are several perspectives on thought experiments. We can speak, for example, with (Cohnitz and Häggqvist 2018; Brun 2018; Stuart 2018), that some thought experiments are "illustrative." That is, they have no evidential role for or against a theory. Instead, they are intended merely to typify the claims of a theory seeking to make them clearer, obvious, and understandable. As an example of this sort of thought experiment, Cohnitz and Häggqvist refer to Plato's Cave, which plays a significant role in illustrating his theory of Forms or knowledge. These experiments lack argumentative value. They do not have the function of affording evidence. Apart from epistemological texts, other examples may include didactical books. In fact, some write textbooks in order to make students acquainted with the entire history of philosophy through thought experiments (Schick & Vaughn 2012). Viewed in this way, thought experiments as such provide no probative force in the context of an epistemological argument. Nevertheless, they may have some function of a different kind. They may have pedagogical or rhetorical significance, for instance.

Also, (Cohnitz and Häggqvist 2018, pp. 407–8) speak about another type of thought experiment, calling them "puzzle cases," which "are cases in which it is either not clear what we should say about the case, or where our pre-theoretic judgment about the case seems incoherent … and they are typically intended to provoke or motivate a theoretical analysis." The proper example of this kind of thought experiment is the so-called trolley case that questions the moral status of the driver of a trolley (Foot 2002, p. 157; Thomson 1985, pp. 1395–6). Some may see that the trolley case is a moral thought experiment, and accordingly, it does not situate in the scope of our interest here. Yet, others may respond by saying that most, not to say all, thought experiments have common epistemic functions, namely, supporting and refuting some claims or the others. Accordingly, that is what situates them in our

[10] I will explain in detail how Plato constructs his thought experiment in the *Theaetetus* in Sect. 1.3.1.1. In the evil demon thought experiment, Descartes imagines an entity that is utterly able to subject the totality of our beliefs to strict skeptical doubt.

[11] Chalmers Zombie thought experiment would be explained in detail in Sect. 1.3.2.4.

scope of interest here.[12] In my view, trolley cases ask us to support or refute some moral claims. So, it puts us in a moral and not an epistemic puzzle. If so, we can say that, just as in an illustrative thought experiment, the epistemic question does not seem to have arisen; the same is the case with puzzle cases in which "we do not know what to say about the hypothetical [moral] case" (Cohnitz and Häggqvist 2018, pp. 408). And obviously, being morally puzzled is not itself an epistemic problem.

I have identified the epistemological use of thought experiment in the argumentative role it plays in the context of an epistemological theory. Yet, with many perspectives on thought experiments, it is then worthwhile to exactly clarify the intended meaning, if there is any, of the term in epistemology. But, since there is no standard definition of it, it is better to provisionally develop our notion of thought experiment covering the most famous cases in the literature. We want to know what a review as such may tell us about the role it actually plays in constructing epistemological accounts.

1.3.1 Thought Experiment in Classical Epistemology

Just as with philosophy, thought experiments and the history of philosophy are interdependent; in epistemology, thought experiments define its history. Epistemologists' usage of their epistemic thought experiments has been common since antiquity. In what follows, I will discuss some representative examples that uncover the roots of this practice in epistemology. Before turning to examples, I would like to mention that, with the intention of tracing the epistemic roots from which this practice arises, I will be following (Rorty 1984)'s genre of rational reconstruction, according to which one can converse with past epistemologists in terms of current epistemology. Doing this also enables us to know about the developments of epistemologists' use of thought experiments, which in turn will help us to elaborate our own sense of the term 'thought experiment.'

1.3.1.1 Plato's *Theaetetus* Case

The dialogue *Theaetetus* deals with the negative aspect of Plato's theory of knowledge. For it does not tell us what knowledge IS but what all knowledge is NOT. For example, he concludes in the dialogue that knowledge is neither sense perception nor true judgment or even a true belief accompanied by an account. His refutation of the third claim was concluded after considering three meanings of the word

[12] In one sense, there is an analogy between "puzzle cases" and epistemic thought experiments. Both have the function of supporting and refuting a theory. Yet, I believe that since we can maintain that, despite both having this common function, they are different from each other. So, I maintain that the analogy does not hold.

'account' or 'explanation': opinion, listing the simplest parts or elements, and locating distinctive characteristics. But they all got rejected, respectively, due to the absence of any criterion that distinguishes a true opinion from a false one.

However, his main argument to refute these three claims –sense perception, true judgment, and true belief accompanied with an account– about knowledge is that all of them yield something which is changeable and universally unacceptable. So, we can negatively conclude that unchangeability, universal acceptability, and indubitability are the necessary features or conditions of true knowledge. And since none of the above-mentioned three claims meets these conditions, they all do not qualify as true knowledge. In order to prove that Theaetetus's view of knowledge as true belief is wrong, Socrates comes up with a thought experiment about the jurors who have a true belief that, intuitively, cannot be counted as knowledge. His main idea is to contrast the juryman's belief and the knowledge of the eye-witness. In Plato's words:

> when a jury is rightly convinced of facts which can be known only by an eye-witness, then, judging by hear-say and accepting a true belief, they are judging without knowledge ... if true belief and knowledge were the same thing, the best of jurymen could never have a correct belief without knowledge (Cornford 1935, p. 141).

However, the dialogues of *Meno* and *Timaeus* deal with the positive aspect of Plato's epistemology. In these two dialogues, his main concern is with what knowledge IS. He gives meaning to the word 'account,' which was not given in *Theaetetus*: to be accompanied with an account is to be established on rational grounds (being unchangeable, universal, and apprehendable by the mind only via reasoning). In doing so, he enables the view of knowledge as true belief accompanied by an account to meet his three necessary conditions of true knowledge. Since then, the concept of knowledge as true belief with an account or as the justified true belief has been associated with Plato.[13]

1.3.1.2 Descartes' Dreamer Case

As a matter of fact, Descartes was never quite far from the charming jungle of thought experiments. In order to establish indubitable knowledge, he made use of a number of thought experiments. For example, in *Meditations on First Philosophy*, he introduces probably the most famous formulation of dream skepticism, which is one of many arguments he offers to challenge the reliability of sensory knowledge. His main idea is that in both dreaming and waking time, one's mind is presented with just the same ideas. There is no standard by means of which one can differentiate between dreaming and waking ideas. Therefore, our claims of knowledge are always dubitable. For instance, he offers a thought experiment or, as he calls it, "a brilliant piece of reasoning" in which he supposes himself to be awake when in fact,

---

[13] As we will see later, this concept of knowledge will be contradicted by what is famously known in recent years as (Gettier 1963)'s cases.

he is asleep. After that, he turns the wheel contrariwise, supposing that he is asleep when in fact, he is not. Here is the thought experiment:

> How often, asleep at night, am I convinced of just such familiar events - that I am here in my dressing-gown, sitting by the fire - when in fact I am lying undressed in bed! Yet at the moment my eyes are certainly wide awake when I look at this piece of paper; I shake my head and it is not asleep; as I stretch out and feel my hand I do so deliberately, and I know what I am doing ... Suppose then that I am dreaming, and that these particulars - that my eyes are open, that I am moving my head and stretching out my hands - are not true (Descartes 1996, p. 13).

This thought experiment shows that we have no precise criterion by virtue of which one can decide whether one is awake or asleep. If so, then our claims of perceptual knowledge are always prone to be accused of being confused by dreams and accordingly being untruthful. True knowledge, for him, is not perceptual. Quite the opposite, true knowledge must be certain, intellectual (i.e., not perceptual), distinct, and obtainable through reasoning alone, exactly as in the example of the *Cogito*, the idea of a triangle, and the like. An attentive reasoning mind is the only security as a means to knowledge.[14] We have seen the view of Plato that sense-perception cannot be regarded as a means to knowledge. Descartes shares this tendency considering reason as superior to sense-perception in yielding knowledge.

1.3.1.3 Husserl's Honeysuckle Case

It is a well-known fact that thought experiment plays a central role in Husserl's phenomenological enterprise of the possibility of knowledge.[15] In his view, knowing is an act. There is unity in the act of knowing. This unit includes two different phenomenological components or acts: one is the intentional meaning or semantic act, and the other is the fulfilling intuition or intuitive act. The latter leads to the former in order to compose an act of knowing that is immediate, conscious, non-inferential evidence. The best way to illustrate what this means is to take a look at the following thought experiment given by Husserl who writes:

> Suppose I believe the honeysuckle in my yard is blooming. All I have here is the performance of an act of meaning: a warranted thought that intends the relevant state of affairs. Now suppose I go into my yard and see the bushes covered in newly blossoming honeysuckle. Where this happens, we experience a descriptively peculiar consciousness of fulfillment: the act of pure meaning, like a goal-seeking intention, finds its fulfillment in the act which renders the matter intuitive (Husserl 1970, p. 694).

The thought experiment well explains the main difference between the above-mentioned two phenomenological acts. The difference is that in an intuitive act, the object (i.e., blossoming honeysuckle) is present before the subject who represents it.

[14] It is worth mentioning that his idea of "attentive mind" plays a crucial role in Sosa's and BonJour's epistemology. For detailed elaboration, see Sect. 1.5.1 in this book.

[15] It is important to note that some recent epistemologists like Chudnoff, for example, found the roots of their phenomenological epistemology of intuition in Husserl's phenomenology.

While in the semantic act, the object (i.e., the honeysuckle is only believed to be blooming) is not present before the subject. The possibility of knowing an object then happens when the fulfilling intuitive act brings the intentional meaning or semantic act with the sensuous presence of the object before the knower who represents it.

If thought experiments play a crucial role in classical epistemology, then the question arises as to whether contemporary epistemology has room for thought experiments. The purpose of the next section is an attempt to put myself in a position to answer this question.

1.3.2 Thought Experiment in Contemporary Epistemology

In our own time, even more than before, epistemology has witnessed essential developments, much of which are grounded on the consequences of thought experiments. So, contemporary epistemology would be a severely impoverished attempt at creativeness without thought experiments. We need only to discuss a short list of the most common thought experiments in contemporary epistemology to make this claim evident. In what follows, I am going to sketch some of the most well-known epistemic thought experiments in contemporary epistemology.

1.3.2.1 Jackson's Fred Case

My first case study is Frank Jackson's "Fred case" thought experiment in which he establishes what is famously known as the 'knowledge argument' which builds pressure on 'physicalism' according to which all mental states can be understood in virtue of physical terms. That is, knowing the physical truths (e.g., knowing all of the physical truths about colors) entails the conceivability of knowing all other types of truth, including phenomenal ones (e.g., what it is like for a person to see color). Contrary to that, Jackson argues that at least mental states like 'qualia' or 'raw feels,' for example, are located out of the range of physicalism. To defend this view, he comes up with the thought experiment of Fred's case in which he asks us to imagine that:

> Fred has better color vision than anyone else on record; he makes every discrimination that anyone has ever made, and moreover he makes one that we cannot even begin to make. Show him a batch of ripe tomatoes and he sorts them into two roughly equal groups and does so with complete consistency. That is, if you blindfold him, shuffle the tomatoes up, and then remove the blindfold and ask him to sort them out again, he sorts them into exactly the same two groups ... What kind of experience does Fred have …? … no amount of physical information about Fred's brain and optical system tells us (Jackson, 1982, pp 128-9).

This thought experiment explains how our having all physical information about the subject Fred won't help us to know everything about his color experience. That is to say that even after we know everything about his brain processes, physiology, history, body, etc., something will still be left out of this knowledge. Put another way, his color experience as a mental state cannot be understood merely in physical terms. If that is the case, then Physicalism is false.

1.3.2.2 Jackson's Mary Case

To reinforce the 'knowledge argument,' Jackson formulates and uses another thought experiment which is famously known as the 'Robot Mary case' in which he asks us to suppose the following scenario. He writes:

> Mary is confined to a black-and-white room, is educated through black-and-white books and through lectures relayed on black-and-white television. In this way she learns everything there is to know about the physical nature of the world … It seems, however, that Mary does not know all there is to know. For when she is let out of the black-and-white room or given a color television, she will learn what it is like to see something red, say (Jackson 1986, p. 291).

One might roughly try to explicate Mary's epistemic progress in the following way. If Mary gets in one way or the other to release herself from the black-and-white room, she will learn about other colors existing in the world. But, if her physical knowledge is complete, she is supposed not to learn anything more. Since she does learn more about the physical world, her previous physical knowledge is not as perfect as physicalism assumed it to be. One may argue that if she learns anything more, her new knowledge may be nonphysical. Even if this is true, again, physicalism is incomplete. For what this proves is that the world might be largely but not entirely physical or interpreted by physical terms.

All in all, the above formulation, however, shows very clearly, Jackson assumes, that the subject's previous physical knowledge is incomplete. That is why, after releasing Mary from her black-and-white room, she will acquire more than all the physical information already known to her. Jackson holds that the same style of scenarios may be deployed for various mental states. Once again, if the argument of her lack of knowledge is true, then the thesis of physicalism is simply incorrect.

1.3.2.3 Searle's Chinese Room Case

Another thought experiment was given by John Searle, which is nowadays well-known as The Chinese Room Argument. Searle's main aim here is to argue against another form of physicalism, namely, the claims of artificial intelligence, according to which an adequately programmed computer can possess the cognitive states just like the human mind. On the contrary, Searle argues that computers and minds seem to be similar in the sense that both can program computational operations. Nevertheless, they are really dissimilar in the sense that computers, unlike minds, cannot have mental processes such as thinking, believing, intending, knowing, and so on. Computers can just simulate the brain's computational operations, but they can never possess the aforesaid internal causal powers a human brain has.

Consequently, the only machine that can think, understand, intend, know, etc., is the human brain. The computer cannot do that because it is a program only and not an intentional brain. If so, then knowing more about computer programs won't help knowing how the human brain produces these fundamental cognitive states. Here is the thought experiment in which Searle imagines himself as a monolingual English speaker and says:

> Suppose that I'm locked in a room and given a large batch of Chinese writing. Suppose furthermore … that I know no Chinese … Now suppose further that after this first batch of Chinese writing I am given a second batch of Chinese script together with a set of rules for correlating the second batch with the first batch. The rules are in English ... They enable me to correlate one set of formal symbols with another set of formal symbols, and all that "formal" means here is that I can identify the symbols entirely by their shapes. Now suppose also that I am given a third batch of Chinese symbols together with some instructions, again in English, that enable me to correlate elements of this third batch with the first two batches, and these rules instruct me how to give back certain Chinese symbols with certain sorts of shapes in response to certain sorts of shapes given me in the third batch. Unknown to me, the people who are giving me all of these symbols call the first batch "a script," they call the second batch a "story," and they call the third batch "questions." Furthermore, they call the symbols I give them back in response to the third batch "answers to the questions," and the set of rules in English that they gave me, they call "the program" ... Suppose also that after a while I get so good at following the instructions for manipulating the Chinese symbols and the programmers get so good at writing the programs that from the external point of view … my answers to the questions are absolutely indistinguishable from those of native Chinese speakers ... But … I produce the answers by manipulating uninterpreted formal symbols … I simply behave like a computer; I perform computational operations on formally specified elements. For the purposes of the Chinese, I am simply an instantiation of the computer program (Searle 1980, pp. 417-8).

However, Searle goes on to wonder whether the manipulation of symbols, however convincing it might be, really comes up to the understanding of the propositions involved. The answer is a clear NO. The best and cleverest manipulation of symbols does not entail knowledge of their meaning. This thought experiment has an epistemological angle to it as it relates to the understanding of propositions and knowledge of meaning, which interacts with the epistemology of linguistic understanding.

1.3.2.4 Chalmers' Zombie Case

In the same vein, David Chalmers argues against another distinct form of physicalism according to which conscious mental facts, especially consciousness, can be reductively knowable in a merely physical sense. That is to say; consciousness is just a physical process with a certain function. However, Chalmers does not agree with this view and argues that physicalism cannot explain the mental property of consciousness.

If physicalism is true, then there should be no possibility for there to be a world that is physically identical to but lacks some components of ours. But a zombie, which is a creature that is physically our duplicate but lacks our conscious experiences, is conceivable or knowable. Therefore, physicalism is wrong. To put it another way, if X is explainable in terms of Q, then Q is not conceivable without X. If consciousness is explainable in terms of physical facts, these physical facts cannot be conceivable without consciousness. But these physical facts are conceivable without consciousness. Therefore, physicalism is wrong. Here is the thought experiment:

> We can imagine that right now I am gazing out the window, experiencing some nice green sensations from seeing the trees outside, having pleasant taste experiences through munching on a chocolate bar, and feeling a dull aching sensation in my right shoulder. What is going on in my zombie twin? He is physically identical to me, and we may as well suppose that he is embedded in an identical environment. He will certainly be identical to me functionally: he will be processing the same sort of information, reacting in a similar way to inputs, with his internal configurations being modified appropriately and with indistinguishable behavior resulting. He will be psychologically identical to me ... He will be perceiving the trees outside, in the functional sense, and tasting the chocolate, in the psychological sense. All of this follows logically from the fact that he is physically identical to me, by virtue of the functional analyses of psychological notions. He will even be "conscious" in the functional senses ... he will be awake, able to report the contents of his internal states, able to focus attention in various places, and so on. It is just that none of this functioning will be accompanied by any real conscious experience (Chalmers 1996, pp. 94-5).

In my view, this argument involves three theses: a modal thesis which is concerned with what is possible (i.e., the possibility of an unconscious zombie), a metaphysical thesis concerned with the nature of our world (i.e., it's being both physical and phenomenal), and an epistemic thesis concerning what is knowable (i.e., the conceivability of a world that has zombies). If it is possible to reduce this thought experiment to an epistemic argument, one can also consider the first two theses as grounds that generate justified belief in the third one as a conclusion.[16]

1.3.2.5 Cohen's Lottery Cases

Stewart Cohen concerns himself with fallibilism, the view that induction is enough ground for obtaining true knowledge. To illustrate, let us say, for example, that a friend of mine came to visit me. When he arrived at my place, he gave me a carry bag on which it was written Vat 69. And it is already known to me that he knows I like this brand of whiskey. Given this, I come to believe that there is a whiskey bottle in the carry bag. My justified belief that there is a whiskey bottle in the carry bag counts as knowledge on the basis of induction. Cohen holds that this ordinary form of knowledge faces a problem that can be shown in what is famously known as 'The lottery cases.' Here, then, are his cases. He first provides us with Case 1:

> Suppose S holds a ticket in a fair lottery with n tickets, where the probability n - 1 /n of S losing is very high. Does S know that his ticket will lose? Although (if n is suitably large) S

---

[16] This depends on whether the thought experiment is a logical argument. Later on, I will argue that this commitment is risky. Instead, I will argue that thought experiments are epistemic arguments.

> has good reasons to believe he will lose, it does not seem right to say that S knows he will lose.

Then, he articulates Cases 2 & 3:

> Suppose S learns from Jones, the person running the lottery, that Jones intends to fix the lottery so S will lose. Does S, then, know that he will lose? Better still, suppose S reads in the paper that another ticket has won. In both of these cases we are inclined to say that S does know that he loses (Cohen 1988, p. 92).

The main idea of these cases is that, in the first case, one cannot know that a particular lottery ticket is going to lose by means of induction from the odds alone. But, in the second and third cases, this is not the case: one can know that a particular lottery ticket is going to lose by dint of the induction from the odds alone. For that reason, the problem of these cases consists in that they amount to contradiction. For, in case 1, the subject does not know that she will lose. But, in cases 2&3, the subject knows that she will lose. If induction is good enough ground for knowledge, the subject in lottery cases would have consistent knowledge. However, it is intuitively clear that she does not. Induction, therefore, issues the wrong verdict in this case.

1.3.2.6 Lehrer's Truetemp Case

Reliabilism in epistemology is an approach that explains the truth of epistemic themes like knowledge, justification, etc., for example, in terms of the reliability of the factor that forms them. For example, something is reliable knowledge if the process that formed it is reliable. In his outstanding book, *Theory of Knowledge*, Keith Lehrer considers two versions of reliabilism. One is an externalist account of knowledge which generally holds that something is reliable knowledge if there is a tie between it and the world. That is, one's belief won't be considered reliable knowledge unless there is a tie between it and what the very belief is about. Another account of knowledge Lehrer considers is internalism, the view that the belief-world connection is NOT sufficient to obtain reliable knowledge. Rather, what is more important is how this connection is made. Lehrer concerns himself with the debate between internalists and externalists in the theory of knowledge. He subjects externalists to criticism by virtue of using thought experiments such as the Truetemp case. Here is the case:

> Suppose a person, whom we shall name Mr. Truetemp, undergoes brain surgery by an experimental surgeon who invents a small device which is both a very accurate thermometer and a computational device capable of generating thoughts. The device, call it a tempucomp, is implanted in Truetemp's head so that the very tip of the device, no larger than the head of pin, sits unnoticed on his scalp and acts as a sensor to transmit information about the temperature to the computational system in his brain. This device, in turn, sends a message to his brain causing him to think of the temperature recorded by the external sensor. Assume that the tempucomp is very reliable, and so his thoughts are correct temperature thoughts. All told, this is a reliable belief-forming process. Now imagine, finally, that he has no idea that the tempucomp has been inserted in his brain, is only slightly puzzled about why he thinks so obsessively about the temperature, but never checks a thermometer to

determine whether these thoughts about the temperature are correct. He accepts them unreflectively, another effect of the tempucomp. Thus, he thinks and accepts that the temperature is 104 degrees. It is. Does he know that it is? Surely not (Lehrer 1990, pp. 163-4).

Among many thought experiments discussed in contemporary epistemological literature, Lehrer's Truetemp case is extensively cited as a paradigmatic case, where the author constructs his thought experiment on the following basis: the subject, Mr. Truetemp, does not know what is the temperature is when he truly says what it is. That is to say; he defends a conception of knowledge according to which the subject must know that the information she has is true over a conception of knowledge according to which it is enough for the subject to obtain information from the world in a reliable way. Hence Lehrer advances internalism over externalism, and his appeal to the Truetemp case is evidence against externalism.

1.3.2.7 Goldman's Fake Barn Case

In recent decades, much has been written about 'fake barn' cases that were introduced into the literature in Alvin Goldman's article "Discrimination and Perceptual Knowledge", where his main aim is to address the question of how the variability in our intuitive assessments can be explained. Suppose you believe that something is true because you see it in a direct manner, with nothing in between. On the basis of these details, if we are asked, we will intuitively say that you know what you perceive. But suppose we come to know that your perceptual evidence misled you to form a false belief. On the basis of these new details, if we are asked, we will intuitively say that you do not know what you perceive.

In the same way, in the fake barn thought experiment, Goldman asks us to imagine Henry and his son getting into their car. Henry drives it in the countryside and has no reason not to identify outside objects as they are. Accordingly, he, without any doubt, identifies to his son an object as a barn, which in fact, it is. Based on this information, if we are asked, we will intuitively say that Henry knows that the object is a barn. But suppose we got to know what is unknown to Henry; namely, the area is full of papier-mâché facsimiles of barns. Yet, when he entered the area, he saw a real barn. But if the object was a facsimile, he would mistakenly identify it as a genuine barn. Now, based on this information, if we are asked, we will intuitively say that Henry does not know that the object is a barn (Goldman 1976, pp. 772–3).

What interests Goldman is how to explain the change in our intuitive assessments from 'Henry knows' to 'Henry does not know'. Without going into too many details of Goldman's view, I shall just say that he suggests that what determines whether our assessment is true or false is the distinguishability or discriminability of the situation. It is worth mentioning that some epistemologists consider Lottery cases, the Truetemp case, Goldman's fake barn country case, and many more as proper examples of the next thought experiment I am going to discuss, namely, the Gettier cases. For all of these cases emphasize the following: there is a gap between being truly justified in believing something on the one hand and getting knowledge about that something on the other hand. Nevertheless, in my view, they are different means intended to achieve different forms of the target.[17]

1.3.2.8 Gettier's Cases

Some philosophers have formulated a tripartite analysis of knowledge as justified true belief. According to this analysis, something is knowledge if and only if it does meet three necessary and jointly sufficient conditions: it must be true, believed to be true, and justifiably believed to be true. In his brief but very influential paper, "Is Justified True Belief Knowledge?", Edmund Gettier revised this long-held view of knowledge by means of scenarios in which the subject has justified true belief without having knowledge. Accordingly, in recent epistemology, the two cases that Gettier sets up are broadly considered important turning points in the history of epistemology: they show how justified true belief can fail to be knowledge.[18] Let us take a look at one of Gettier's two cases.

The first example Gettier comes up with has to do with Jones and Smith, who have applied for a particular job. Suppose that the boss told Smith that Jones would be offered the job. Suppose also that Smith has already checked and got to know that Jones has ten coins in his pocket. With the given evidence, Smith has a solid justification for believing the following proposition is true: the one who will get the job has ten coins in his pocket. In this case, if Jones got the job, we would say that Smith knew that Jones would get the job. But suppose further that Smith, in the end, was selected and that, unknown to him, he himself has ten coins in his pocket. In this case, we have to say that although he is justified in believing the previous proposition, Smith does not know that it is true. In fact, his belief that it is true is just a kind of luck (Gettier 1963).

Gettier's two examples have been hotly debated in recent literature. It was often argued that they successfully function as a serious challenge to the tripartite analysis of knowledge. Thus, they had a dramatic impact on epistemology. Their evidentiary status designates and secures them. In the 10 coins-case, which was just presented above, for example, it is obvious that Smith does not know the proposition 'the one who will get the job has ten coins in his pocket' is true. In fact, he was very lucky to believe it was true.

[17] In one sense, there is an analogy between these cases. They share the same objective (i.e., justified true belief does not qualify as knowledge). Yet, they are different in two senses. One is that different cases of them produce evidence against different types of the theory of knowledge as justified true belief. For example, for Cohen's Lottery cases, inductively justified true belief is not knowledge. While for Lehrer's Truetemp case, lucky justified true belief is not knowledge. Another is that they produce different evidences against the very theory. That is, some of them are constructive, others are destructive, etc. This latter point will be clearer in Sect. 1.4.1, where I will discuss both characterizations of thought experiment and their epistemic function.

[18] It is worth noting that Gettier's cases are not purported to provide a definition of what knowledge is. They challenge the sufficiency of justified true belief description as a definitional analysis of knowledge.

1.3.2.9 Kripke's Gödel–Schmidt Case[19]

In his book, *Naming and Necessity*, Saul Kripke argues against the descriptivist theory of reference held by Frege, Russell, Searle, and many others, whose main concern is the relation between language and the world. According to this theory, proper names with definite descriptions associated with them refer to the object that meets such descriptions. Attacking this view, Kripke provides several arguments, such as the epistemological argument that goes as follows. The proposition 'if proper name exists, then proper names are definite descriptions' is not known a priori. Or it is not known a priori as true. If it were otherwise, we would be wrong. But descriptivism entails otherwise. So, descriptivism is fallacious.

The book is well-supplied with thought experiments that support this argument. One of them is the well-known Gödel-Schmidt example in which he asks us to imagine the proper name "Gödel," which is associated with the definite description as *the author of the incompleteness theorems*. But he asks us to suppose further that Gödel has stolen the script of the theorem from its real author Schmidt who passed away immediately after coming up with the theorem. Gödel published the work in his name, and this is how since then, people have associate with his name the definite description "the author of the incompleteness theorems" (Kripke 1980, pp. 83–4).

Now, whether or not any of such things indeed occurred, the main thing is that we do not know it a prior, although we know a priori that the author of the incompleteness theorems wrote them. Moreover, if the descriptivist theory is true, we should refer to Schmidt when we use the name, Gödel. But we do not do that. So, the proposition 'Gödel wrote the incompleteness theorems' is different from the proposition 'the author of the incompleteness theorems wrote the incompleteness theorems.' The latter is knowable a priori, but the former is not. What is more, "Gödel" is a proper name. The description associated with the name is the proposition 'the man who wrote the incompleteness theorems.' The proper name and description do not refer to Schmidt. Therefore, descriptivism is wrong.

1.3.2.10 BonJour's Example of Norman the Clairvoyant

In the first part of his book, *The Structure of Empirical Knowledge*, Laurence BonJour's main goal is to show the inadequacy of foundationalism in providing a satisfactory explanation of empirical justification. For foundationalists like Descartes, Armstrong, and others, empirical beliefs are fundamental or basic: they are justified, and their justification does not depend on other empirical beliefs that themselves are in need of justification. Not so, argues BonJour, saying that "there is no way for the foundationalist's allegedly basic empirical beliefs to be genuinely

[19] It is worth mentioning that in his book, Kripke uses this case and many others in various contexts of arguing against descriptivism. Here I am concerned with his usage of it in the context of what he calls as the epistemological argument.

justified for the believer in question without that justification itself depending on further empirical beliefs which are themselves in need of justification" (BonJour 1985, p. 84). If so, however, foundationalism clearly entails further problems.

BonJour's main problem with empirical foundationalism is that it amounts to the epistemic regress problem. This serious and fundamental problem commonly takes the following form: "to know any proposition P we must know a proposition Q that provides evidence for P, but this requires an endless regress of known propositions—a circle or an infinite regress—so, since we cannot acquire knowledge by means of such regresses, knowledge is impossible" (Cling 2014, p. 162). At the opposite extreme, the externalist version of foundationalism responds to the epistemic regress problem by claiming that "the justification of a basic belief depends on an external relation between the believer (and his belief), on the one hand, and the world, on the other, specifically a lawlike connection" (BonJour 1985, p. 35). In other words, contrary to internalism, externalism holds that justification is an entirely external/nonmental matter and that it is NOT always possible to know what justifies your belief when it is justified. That is to say, even when your belief is justified, you may not know what justifies it simply because its justifier may not always be accessible to you.

BonJour rejects the externalist's response and presents four clairvoyant cases to show the major problems with externalism. Due to our purpose here, it is enough to focus only on the case of Norman, which can be described as follows. Norman is a completely reliable clairvoyant, in certain conditions which can be usually obtained, with respect to certain kinds of subject matter. He possesses no evidence or reason of any kind for or against the general possibility of such a cognitive power. One day Norman comes to believe that the President is in New York City, though he has no evidence either for or against this belief. In fact, the belief is true. Is Norman epistemically justified in believing that the President is in New York City so that his belief is an instance of knowledge? (BonJour 1985, p. 41).

According to the externalist view, we must say that he is; simply because, according to externalism, "a person's beliefs might be epistemically justified merely in virtue of facts or relations that are external to his subjective conception" (BonJour 1985, p. 36). If so, if what justifies one's belief is just one's external facts or relations and not one's beliefs and other cognitive states, the regress of justification stops occurring, and the epistemic regress problem is resolved, externalists hold. In contrast, BonJour thinks that the problem at issue, however, is obviously more serious. He rejects the externalist response because it reveals a poor understanding of the epistemic regress problem that came into view in terms of justification rather than knowledge. Additionally, it seems safe to say that the externalist response is hopelessly irrelevant to the epistemic regress problem. Despite the problem of changing the subject, the thought experiment shows how the subject is epistemically unjustified depending on her mere external facts or relations and not on her internal beliefs, and accordingly, such belief is not an instance of knowledge.

The goal of this section is limited to making some observations about appealing to epistemic thought experiments in contemporary epistemology. If we are asked to come up with further paradigms, many more interesting pieces of thought experiments are available. Examples may include Burge's arthritis case, Cohen's airport case, and many others.[20] Such examples are not hard to explain, but the cases presented so far offer an adequate background for our upcoming considerations: what are the common characteristics of these thought experiments, their structsure, and function?

1.4 Commentaries on Thought Experiments in Epistemology

In previous sections, we have seen how thought experiments dominate epistemology. In particular, I have sketched some of the best-known thought experiments that play a significant methodological role in epistemology. Although thought experiments are quite common in many areas, what makes thought experiment in epistemology different is that epistemologists make it the object of study itself. They concern themselves with questions of its common nature, features, etc. In the current debate over the nature of thought experiment in analytic epistemology, there are many characterizations of thought experiment. Examples of these characterizations could be empiricism, according to which thought experiment is a sub-argument (Brendel 2018), rationalism, which considers it to be a stimulation of rational insight (Grundmann 2018); naturalism, which looks at it as stimulation of cognitivists' mental models (Nersessian 2018), Kantianism that treats thought experiment as a continuous function with a real one or/and a condition for a possible one (Buzzoni 2018), and phenomenologism that deals with it as a first-person anticipation pump (Wiltsche 2018).

In what follows, I shall be attempting to characterize thought experiments. To that end, I shall try to explore its structure and function. To do so, I will borrow and discuss some common portrayals offered by some contemporary philosophers who are interested in this subject. I shall make use of these depictions in order to draw my own picture of the issue at hand. By doing this, I can proceed to create a safer pathway for carrying out the purpose I have tasked myself to fulfill in the next step, namely, the nature and evidential role of epistemic intuition.

[20] In brief, Tyler Burge's thought experiment concerns Oscar, who has an imperfect understanding of the concept of 'arthritis.' Accordingly, he falsely thinks that he has arthritis in his thighs. Oscar got corrected and was convinced by the medical definition of arthritis given by his doctor. While Stewart Cohen's airport case concerns Mary and John, who are at Louisiana airport and want to fly to New York by a certain flight. They want to know if it has a stopover in Chicago. They overhear from a passenger, Smith, who tells someone after checking the plane itinerary, that it has. Mary rightly suspects this. Hence, they change the schedule agreeing that Smith does not really know what he uttered.

1.4.1 *Characterizations of Thought Experiment and Their Epistemic Function*

My aim in this section is to give proof or evidence that clearly shows the variety of accounts proposed to explain the thought experiment. The point is also to show that thought experiment has a certain epistemological depth and thereby attempt to elaborate the basic ideas of the ongoing discourse in the philosophical literature into a more comprehensive account of epistemic thought experiments.

The standard characterization of thought experiments is that they serve to describe a scenario of a postulated development of events. If so, the question now is, do they weave concrete or possible scenario? Some hold that a thought experiment is a scenario woven from some particulars of the real world (Hintikka and Hintikka 1989; Häggqvist 1996). On the other hand, others argue that it is woven from some psychological particulars, such as imagination, for instance. Each of these two positions is partly right: I think that thought experiments may be woven from both sources, the real world and the use of the imagination. For example, Lehrer's Truetemp thought experiment makes us think about whether we should count Truetemp's thoughts caused by an imaginary device implanted in his head as reliable knowledge. Whatever this imaginary device is, it creates the very frame of this thought experiment, and accordingly, it is essential, particular to the relevant thought experiment. Moreover, the same thought experiment is, to some extent, woven from real-life particulars. For instance, in addition to the above-mentioned imaginary device, Truetemp's thought experiment also includes concrete particulars such as the surgeon, thermometer, and temperature, which help us to continue with the thought experiment and consider it relevant. These two kinds of components lead us to discuss the following two characteristics of thought experiments: argument and counterfactuality.

To my understanding of the word 'argument', it means reason(s) or evidence(s) produced for the claim that this or that theory is justified or unjustified.[21] For example, Oscar has pain in his thighs, and he knows there is a disease called 'arthritis' which causes painful inflammation and stiffness of some part of the body. This justifies his thinking that he has arthritis in his thighs. But Oscar got to know the medical definition of arthritis given by his doctor. For that reason, Oscar's previous understanding of the concept of 'arthritis' is imperfect. If my understanding is correct, then I think one important lesson, maybe among many, is to be learned, namely, thought experiments *are* arguments. Among proponents of the argument-based view of thought experiments are (Norton 1991, 1993, 1996); (Rescher 1991); (Irvine 1991), and (Forge 1991). In other words, a thought experiment offers an argument for or against a target theory. When it is against, it takes the form of a counterexample. In fact, there is "a usual practice of analytic philosophy of giving

[21] This point will be discussed again, and what is meant here by the word 'argument' will become clearer when I argue in the third chapter that thought experiments provide arguments in an epistemic and not a logical sense.

counterexamples with the practice of constructing thought experiments" (Engel 2011, p. 157). A thought experiment is a counterexample when it has an explicitly negative implication that disproves an assertion made by a claim of an already-standing theory. For example, suppose there is a theory that says every A is B. If you can find a thought experiment in which at least one A is not B, then that thought experiment will be considered as a counterexample against the already-standing theory. Thought experiment, as a counterexample, commonly works as follows: it shows that something is the case and makes use of it in an argument against a theory. Smith, the subject of Smith-Jones Gettier's example, who has justified true belief but no knowledge, was used to provide such a refuting argument against the claim that knowledge is justified true belief. Or, Mary, the subject of Jackson's example who got to know more about colors after she was freed from the black-and-white room in which she was raised, was used to provide such a refuting argument against physicalism, according to which all truths are physical ones.

But it is argued that we cannot say that thought experiments are arguments for many reasons: first, thought experiments and arguments are two different entities (Brendel 2018). Otherwise, a thought experiment should be called explicitly an argument, not a thought experiment. The problem with this objection is that it overlooks the fact that a thought experiment may be implicitly argument, communicating analogous thoughts. Second, if by 'argument,' it is meant logical argument, this reduces the thought experiment to a proposition. And thought experiments are more than logical propositions; they include some psychological elements like imagination, for example (Cooper 2005). Here I have to maintain that I am not following Norton's and his colleagues' usage of the term 'argument.' They use 'argument' in a logical sense. While I use the very term in an epistemic sense. An epistemic argument includes reasons, grounds, or pieces of evidence that produce justified or unjustified beliefs and exists entirely without propositions that are sensitive to the truth conditions under which the given proposition is true or false. Thus, this strategy, I think, enables my usage of the term to avoid the charge of reducing the concept of 'thought experiment' to the notion of 'proposition.'

Third, arguments are not opaque, but we usually feel uncertain because of the opacity of the thought experiment. Such opacity causes disagreement among philosophers on the validity of thought experiments. If a thought experiment is obviously an argument, then there should not be a disagreement among philosophers on its validity; but disagreement is there (Häggqvist 2019). However, it is imagination that causes the opaqueness of some details of thought experiments, as in Chalmers' world of Zombies. The reason behind the unclarity of worlds as such is that they are not actual worlds. That is, some thought experiments present a scenario that is called in contemporary epistemology as involving possible worlds, i.e., total ways that the world might have been. For example, Gettier's example is a possible world for the conception of knowledge as justified true belief. In Gettier's possible world, it is possible for one to have justified true belief without knowledge.

Furthermore, possible worlds take, in some philosophical writings, the form of 'what if' a certain scenario happened differently. For example, in the actual world, one cannot always predict the temperature correctly, but 'what if' it happened that someone called Truetemp could always predict the temperature degree correctly, as in Lehrer's scenario of the Truetemp case? This is a topic where possible worlds take the form of a 'counterfactual,' and accordingly, the possibility is understood in terms of models relative to the modalities (e.g., possibility, impossibility, necessity, contingency, etc.). In this context, (Sorensen 1992) characterizes thought experiments as "alethic refuters," and (Dohrn 2018) claims that there is a consensus on the idea that thought experiment is counterfactual. However, this claim won't withstand Kripke's argument that I adopt. According to it, the differences, if any, between possible and actual worlds are minor.

An opponent of thought experiment as argument may object that it is important to distinguish between three sorts of possibility, logical, metaphysical, and epistemic. A thought experiment is logically possible when it includes what is consistent with the fundamental laws of logic. For example, whether or not anything of Gödel–Schmidt scenario indeed happened, it is logically possible—there is no latent inconsistency in supposing Gödel has stolen the script of the theorem from its real author Schmidt. Gödel published the work in his name, and this is how people associate with his name the definite description of *the author of the incompleteness theorems*. Those who do not admit the view of thought experiment as argument take 'argument' to be understood as having a logical form. Or in (Stuart 2016)'s words, they focus on "the logic of thought experiments." For them, epistemologists make an appeal to thought experiment as a counterfactual, a form of 'what if' or a form of 'if … then'.

But the counterfactual conditional is subject to a serious problem: it violates the property of transitivity that constitutes the syllogistic argument (Luisa & Chiara 1992). Consider the following example. Premise 1: If Hilary Putnam were a scientist, then he would not be wrong. Premise 2: If Hilary Putnam were not wrong, then water would not be H_2O. Conclusion: If Hilary Putnam were a scientist, then water would not be H_2O. In the framework of possible worlds, one can imagine a possible world where Putnam is a successful scientist who discovered that water is xxx. The transitive property is violated because our assertion of the second possible world or second premise does not necessarily include what we assert by the first possible world or first premise. As a result, two premises may be true, but the conclusion may be false. 'Water is not H_2O' breaks down the laws of nature. What all this shows is that it is better not to take thought experiment as a counterfactual argument draped in the logic of modality. Thought experiments would lose much of their force if that happened.

This brings us to the second sort of possibility, metaphysical possibility. For it might be argued that if I am right about the failure of transitivity and if the counterfactual conditional breaks laws of nature, then (Ghins 2007, p. 127) is right: "the truth of scientific laws is based on a metaphysics of nature." Inasmuch as thought experiment includes what is inconsistent with the fundamental laws of nature, it speaks of what is metaphysically possible, perhaps as in the example of zombies. That is to say, the laws of nature and science as we know them now do not admit as acceptable the idea that zombie-like creatures, which are physically identical to human beings but have no consciousness, can exist. Recall that a few decades ago, it was odd to claim that the laws of nature and science allow human beings to go to the moon. In my view, Zombiehood is metaphysically possible in the way it was, a few decades, ago odd to claim that the laws of nature and science allow human beings to go to the moon. The example of going to the moon shows that it is risky to understand a thought experiment as a metaphysical possibility simply because the output of doing so may be making the thought experiment be taken under a false sense of possibility.

The third sense of possibility is the epistemic possibility, according to which what is possible is relative to the thought experimenter's epistemic state—what is possible for all she (the thought experimenter) knows (Dohrn 2018, p. 373). For instance, for all Gettier (the thought experimenter) knows, justified true belief does not entail knowledge. This understanding of epistemic possibility is compatible with my above-mentioned understanding of argument as a list of ground(s), reason(s), or evidence(s) that justify each other, producing justifiedness or unjustifiedness for the claim that this or that theory is justified or unjustified. With this in mind, the thought experiment is an argument that is innocent of all charges of metaphysical and logical commitments. It is grounded in the experimenter's limited knowledge of the world, her epistemic commitment.

At any rate, a thought experiment as an argument is directed at destroying, establishing, or simultaneously destroying and establishing a theory. One way to clarify this point is to use Brown's (1986, 1991) taxonomy. Brown distinguishes between three ways in which a thought experiment can be directed at a theory. First, thought experiments may be destructive. A thought experiment is destructive when it aims to destroy an old theory and generates no new one. For example, Jackson's Fred and Mary cases, Searle's Chinese Room case, and Chalmers' Zombie case are aimed at refutation of different forms of physicalism. More examples of a destructive thought experiment may include Gettier's refutation of the conception of knowledge as justified true belief and Cohen's refutation of fallibilism.

Second, the thought experiment may be constructive. A thought experiment is constructive when it aims to establish evidence for a theory. Examples of a constructive thought experiment may include the Truetemp case, in which Lehrer establishes evidence for internalism, and the Fake Barn Country case, in which Goldman establishes evidence for the theory that what determines whether our assessment is true or false is the distinguishability or discriminability of the situation.

Third, the thought experiment may be Platonic. That is, it is simultaneously destructive and constructive. It aims at both an act of destroying an old theory and an act of establishing evidence for a theory at the same time. As an example of a Platonic thought experiment, Brown gives Galileo's Falling Stones thought experiment, in which Galileo proves Aristotle's view, according to which weighty objects fall quicker than light ones to be wrong and replaces it with his own view according to which all such objects fall at the same rate. This thought experiment is platonic because it refutes Aristotle's theory and at once establishes a piece of evidence for an alternative theory. In the same way, Kripke's Gödel-Schmidt case simultaneously refutes the descriptive theory of names and suggests the causal-historical theory of reference. Analogously, BonJour's case of Norman the clairvoyant, at the same time, poses a serious threat to the externalist version of foundationalism and motivates the internalist account of justification.

This leads us to a character of thought experiments called by (Brendel 2004) as "the informativeness of thought experiments" (p. 93). That is, in virtue of thought experiments, "we learn something new about theories and concepts" which already exist. Thought experiments, as such then, have crucial epistemic significance. We can say, for example, that by virtue of Galileo's Falling Stones thought experiment, we learn something new about the world; namely, Aristotle's view is wrong, and all objects fall at the same speed.

The focus has been so far on how a thought experiment as an argument is directed at a theory. However, this directedness is intentional. More precisely, the scenario is framed in the context of the experimenter's cognitive intention(s). That is, the thought experiment is a result of the experimenter-intentional-directedness. In other words, "it is correct to say that someone [i.e., the experimenter] did something intentionally under a certain description" (Ludwig 2007, p 129). What a thought experiment shows is intentional content (Burge 1982). For example, Gettier had a certain intention (i.e., him being interested in the question about knowledge) that made him formulate his cases as they are: the experimenter –intentional-directedness frames the details of the scenario. That is why he questioned the usual conception of knowledge, proving its weakness by giving two examples in which he came to conclude that justified true belief does not qualify as knowledge.

1.4.2 The Structures of Thought Experiment and Their Epistemic Function

One may be motivated by the view that the epistemic function of a thought experiment cannot be completely captured in terms of its distinctive features, but rather, its structure is also required to be accounted for. I take it that, seen in this light, it is also important that one investigates the structure of the concept one is concerned with. In what follows, I elaborate on the most basic ideas of recognized accounts of the structure of thought experiments in the literature and employ them to continue constructing a more sophisticated account of epistemic thought experiments.

Every thought experiment depends on an initial assumption whose function is directly relevant to drawing a conclusion (Meynell 2014; Myers 1986; Norton 1996; Brendel 2004). For example, the assumption of Gödel-Schmidt's example is that the only description associated with the proper name "Gödel" is the proposition 'the man who wrote the incompleteness theorems.' Having made this assumption, Kripke has come to two kinds of conclusions. One is epistemic: we cannot know the proposition 'Gödel wrote the incompleteness theorems' a priori. Another is semantic: the descriptivist theory of reference according to which proper names (e.g., "Gödel" in this case) with the definite descriptions associated with them (e.g., the author of incompleteness theorems in this case) refer to the object that meets such descriptions (e.g., Schmidt in this case) is wrong.

The same story goes for BonJour's example of Norman, whose initial assumption is that the only evidence or justification, if any, Norman has for his belief that 'the President is in New York City' is luck. Having made this assumption, BonJour has come to the conclusion that such belief is not an instance of knowledge, and accordingly, externalism does not hold.

Along the same line, (Ierodiakonou 2011) provides us with an analogous characterization of thought experiment. In her view, thought experiments are suppositional stories beginning with a fictitious assumption that puts one before two different potentials. These two options need to be investigated in order for one to know which of them one should follow to draw a clear conclusion. For this reason, one's approach to a thought experiment should direct her attention towards scrutinizing the assumption made to know its function in conclusion.

Idealization is another essential ingredient of thought experiment. By the word 'idealization,' I mean that the description of the relevant thought experiment is perfect. For instance, Chalmers' Zombie as an anti-physicalist thought experiment would not be successful without his strategy of idealizing the description of both the conceivability and possibility theses: the description of a conceivable world of Zombie, and the description of a possible world that satisfies to be exactly like ours with one exception that consciousness is not there. If so, if idealization is what makes a thought experiment exists, idealization must play a central role in enabling us to distinguish between successful and unsuccessful thought experiment. In (Norton 2018)'s words, epistemologists usually focus on successful thought experiments. But some thought experiments fail because they lack idealization. A thought experiment can be idealized if the experimenter has enough freedom to consider it to be ideal. Now, when can we say that an idealization is justified? Idealization is justified if it is informative and its effects are knowable (Reiss 2018). Meeting those two conditions enable a thought experiment to acquire its epistemological power successfully. Without idealization just described, Chalmers would not be able to reach his conclusion that disproves the idea that consciousness can be explained by physical terms only. What is left is the question of whether idealization makes thought experiment similar to computer simulation (Lenhard 2018). I believe with Lenhard that the answer is no. For thought experiment is completely accessible to epistemic assessment but computer simulation is not.

Nevertheless, Lenhard holds that both thought experiment and computer simulation emerge from a common root, imagination. This leads us to another conspicuous component of thought experiment: imagination. In the process of conducting a thought experiment, the experimenter imaginatively interacts with images. These images are pillars which instruct her to imagine certain scenarios in a particular way (Meynell 2018). Consequently, these pillars play a needed role in justifying the conclusion of the imagined scenario. That is, they provide the experimenter with a justification for its epistemic success. If so, imagination plays an essential role in warranting the epistemic success of a thought experiment. It is what enables the relevant thought experiment or invented narrative to increase our knowledge about the world through assembling unjointed cognitive resources (Davies 2018). Cohen's lottery cases, for example, increase our knowledge about the nature of knowledge. They teach us that induction is not enough ground for obtaining true knowledge.

Last and not least, it is often argued that intuition is the most salient component of thought experiment. It is also said that this component also plays the role of evidence for the thought experiment that involves it. Now, if epistemology, especially the contemporary variety of it, depends on thought experiments that, in turn, count on intuition, then intuition plays a crucial role in contemporary epistemology. Here is how Hilary Kornblith introduces the assumption that epistemologists, as a matter of fact, rely widely on intuitions as evidence in their theorizing: "George Bealer does it. Roderick Chisholm does it a lot. Most philosophers do it openly and unapologetically, and the rest arguably do it too, although some of them would deny it. What they all do is appeal to intuitions in constructing, shaping, and refining their philosophical views" (Kornblith 1998, p. 129). Alvin Goldman, another proponent of intuition-based epistemology, writes: "One thing that distinguishes philosophical methodology from the methodology of the sciences is its extensive and avowed reliance on intuition" (Goldman 2007, p. 1). Thus, reliance on intuition taken as a philosophical method is what differentiates philosophy from science.

However, the experimental philosophers' version of this normative claim may be more direct and radical. (Weinberg 2007), for example, tells us that "analytic philosophy without intuitions just wouldn't be analytic philosophy" and "that intuitions have become not just one of philosophy's tools but part of its subject matter as well" (p. 318). As I see it, contemporary epistemologists consider intuition-based epistemology as plain truth. What evidence Jackson and Searle, for instance, provide is only intuition that our having all physical information about the subject Fred won't help us to know everything about his color experience, that Mary's previous physical knowledge is incomplete, and that any computer lacks cognitive states.

Given the centrality of thought experiments in epistemological inquiry, the better one's account of thought experiment as evidence is, the better understanding of the nature of epistemological methodology one gains. Upon this motivation, (Stich and Tobia 2018) differentiate between two kinds of intuitions used by epistemologists as evidence in thought experiments. One is used as evidence for mental entities, such as the content of concepts, the justification for tacit theories, etc. Another is used as evidence for extra mental entities such as universals, modal truths, etc. Stich and Tobia argue that while we have no good reason to defend the reliability of the latter, experimental philosophy (i.e., survey-based methods) shows the unreliability of the former. This unfaithfulness of intuition was argued for mainly on the basis of the fact that the strength of intuition varies according to factors irrelevant to the topic of the relevant thought experiment itself.

However, the point of variability in intuitions was defended by (Cohnitz and Häggqvist 2018). Like (McAllister 2018)'s historicist approach, according to which thought experiment is historical attainment and, accordingly, its historical and cultural contexts underlay its epistemic status, the proposed argument argues that thought experiments are best explained as follows. Different factors formulate a dialectical context for thought experiment that involves intuitive modal judgment.

Different sub-epistemological fields require different things from these intuitions. If so, either several epistemological accounts of intuition are necessary, or we have to consider the dialectical context of thought experiments that involve these intuitions in order to formulate a complete epistemological account.

Another motivation for this variation comes from (Bokulich and Frappier 2018). According to them, the thought experiment is changeable over time, as in Lucretius' version of Archytas' thought experiment.[22] In the same vein, (Goffi and Roux 2018) propose a dialectical account of thought experiment according to which thought experiment is a cognitive, contextual and counterfactual scenario. Contrary to that, it can be argued that these methods can show the variability of intuitions without being able to cast any serious charge on their reliability (Ludwig 2018). The variability of intuitions and their reliability are two different things. Also, the invariability of intuitions does not entail their reliability. Therefore, experimental philosophical results are unwarranted.[23]

We have come to the conclusion that an epistemic thought experiment is an imaginative, intentional, and concrete or possible scenario that takes a narrative form of a postulated development of events and can be best analyzed as an argument of the sort defined. This argument is accompanied by a crucial epistemic significance of informativeness relative to the thought experimenter's epistemic state. And the such argument is directed at destroying, establishing, or simultaneously destroying and establishing a theory. It does mainly and really so in virtue of the help offered by two unique components: an idealized assumption and an intuition arises from it, producing either justifiedness or unjustifiedness of the pertinent theory.[24]

But then again, the main epistemological concern is the same, the one I raised right at the beginning: what is the nature of intuition. In recent years, this concern has been increasingly and repeatedly discussed. In order to locate my small contribution in responding to this urgent query within the scope of these ongoing debates, intuition will be my next concern beginning with its emergence in epistemology.

---

[22] Archytas' thought experiment was later taken over and modified by many philosophers like the Roman philosopher Lucretius, for example. By imagining the man who tries to throw a spear stands at the outer edge of the world, Lucretius supported two doctrines: the plurality of the worlds and the doctrine that the world has no fixed limits.

[23] It is worth noting that here I neither defend nor deny any of these accounts. For regardless of the rightness or wrongness of any of them, all that concerns me here is to show how intuition arises as a genuine component of epistemic thought experiment.

[24] For detailed elaboration about how the abilities and limits of thought experiment can be traced to the fact that intuition arises from it, see (Ichikawa & Jarvis 2009).

1.5 Intuition in Epistemology

While the question of the nature of intuition is not only young but highly contentious in epistemology, philosophers' usage of intuition in their epistemic thought experiments has been common since antiquities. For since intuitions arise from thought experiments and philosophers have used thought experiments since the term 'philosophy' was coined, then intuition's usage in philosophy is as old as the subject itself. If this is the case, then it seems plausible to initially look for a historical explanation of the term 'intuition'. In what follows, I would like to focus on the history of the word 'intuition,' its origin, what it stands for, and how its meaning has changed over time.

1.5.1 Epistemic Intuition in Classical Epistemology[25]

There is a long tradition of the activity of philosophizing that produces an inclination in many philosophers to set out to write philosophy in the way in which intuition plays a key role in epistemological inquiry to the extent that their epistemological writings are marked by heavy doses of intuition. Nevertheless, epistemologists have never shown an interest in developing an overall definition of the term 'intuition.' Accordingly, it is better to start with the etymology of the word 'intuition.' The primary conception of intuition is ancient. According to (Lee 1989, p. 2), the word can be traced back to the Latin verb '*intueri,*' which is roughly equivalent to obtaining knowledge from an internal source. In Greek philosophy, especially Plato's and Aristotle's systems, the term was used as one of the cognates of the terms '*nous*' or '*noesis.*' That is, it referred to a "direct intellectual cognition" with "an intersubjective communicability and verifiability" (Piętka 2015, p. 24). This can be easily shown with reference to the aforementioned Theaetetus thought experiment. According to the way by which contemporary philosophers analyze thought experiment, Plato in the *Theaetetus* relies on the intuition that the jurors lack knowledge. This intuition is a direct cognition since Plato did not argue for it.[26] And Socrates got it intersubjectively verified by asking Theaetetus whether he was right in comparing the juryman's belief and the knowledge of the eye-witness, who responded directly by saying, "certainly."

In modern western philosophy, "it was, in fact, Descartes who had first employed the word intuition to mean "the conception of a pure and attentive mind, which is so easy and definite that no doubt about what we are understanding can remain" – a

[25] It would take us too long to completely explain the notion of intuition in classical epistemology. A few words, however, should be said about some of the most interesting issues in this regard.

[26] In one sense which is widespread in contemporary epistemology, intuition is frequently deemed to be fairly spontaneous, immediate, and unreflective knowledge (Goldman & Pust 1998, p. 179; Cohnitz & Häggqvist 2009; Nado 2015; Hannon 2018).

usage that he acknowledged to be a novel use of the Latin, *intueri*" (Meehan 2014, p. 94). Applying this meaning of the term 'intuition' to his above-discussed thought experiment of the dreamer, he relies on the intuition that the dreamer's belief is typically false. For, in dreams, one might touch objects which do not exist. And since we have no precise criterion by means of which one can decide whether one is awake or asleep, our claims of knowledge are no more able to be relied on as true. That is why he considers dreams to be a serious threat to our claims of knowledge.

Moreover, it is worth mentioning that his idea of an "attentive mind" plays a crucial role in Sosa's and BonJour's defense of the evidentiality of intuition.[27] On Sosa's view, for example, intuitions are products of mental competence. Although they are fallible, intuitions are reliable because they are competent. Their fallibility, when they are fallible, is due to the subject's inattentiveness; they are correctable by paying more attention. When intuition is false, its falsifiability is not relative to the competence itself. The same story is not true in the case of other faculties of cognition, such as perception and introspection, where the frailty of intuition is relative to the system itself.

In the same way, BonJour's view of both the source and evidential status of intuition is very close to Sosa's. Just as in Sosa's view, intuition is a product of mental competence, in BonJour's view, it is a product of a "cognitive state." And just as in (Sosa 2007a, p. 59)'s view, intuition may be fallible due to "performance errors chargeable against the subject, by contrast with deliverances of a competence;" yet, the fallibility of intuitions can be avoided by exercising more care and attention; in (BonJour 1985, p. 208)'s view, "a mistake in a priori intuitive apprehension can only result from some sort of confusion or unclarity;" yet, the frailty of intuition "can always be corrected by exercising greater care" (Tidman 1996, p. 164). What is more, this Cartesian rational view of intuition- derived from an attentive mind- as the source of knowledge is a common thought with Bealer, who also sees rational intuition as a source of knowledge, defining it as intellectual seeming.[28]

In his book *Ideas*, Husserl identifies intuition simply as "intellectual seeing" (Husserl 1983, p. 32). According to him, this intellectual seeing enables us to be conscious of the essence of objects (Husserl 1983, pp. 9–10). Further, in *Logical Investigations*, he draws a clear distinction between signitive and intuitive intentions. He writes that "a signitive intention merely points to its object, an intuitive intention gives it 'presence,' in the pregnant sense of the word, it imports something of the fullness of the object itself" (Husserl 2001, p. 233). To explain that, he utilizes some terms such as 'matter' by which he means intentional content. The best way to illustrate the meaning of this is to take a look into his above-discussed thought experiment of the honeysuckle, in which he distinguishes, as we have seen, between intuitive and semantic acts. In intuitive act = intuitive (full) intention, the object (i.e., blossoming honeysuckle) bodily presents before the subject who represents it, while in semantic act = signitive intention, the object (i.e., the honeysuckle believed

[27] Sosa's view of the attentive mind will be discussed again in detail later in this book.

[28] Bealer's view of rational intuition will be discussed again in detail later in this book.

to be blooming) does not present itself before the subject; the object is just signified without presence.

As a related matter, some recent theorists about intuition like Chudnoff, for example, start from what Husserl calls 'matter' as "a dependent part of the psychological event that represents the relevant object (in this case, the honeysuckle itself)".[29] Chudnoff believes that intuition is true just in virtue of this phenomenological representation. His view that intuition is an experience and analogous to perception can then finds its roots in Husserl's phenomenology. The danger is that if so, if "every part of the expansive body of human knowledge, according to Husserl's view, finds its origins in some act of fulfillment, be it in my own personal psychological history or in that of some other individual or group of individuals that has been handed down in scientific practice and culture;" then recent empirical theorists' thesis about intuition according to which intuition varies according to factors irrelevant to its truth such as culture, social status, or so on is true (Kidd 2014, p. 132–3).

For Husserl, however, there are minimally three sorts of intuition with three different kinds of objects.[30] They are: (1) perceptual intuition qualifies as inadequate knowledge because its content is mere appearance; (2) categorial intuition synthesizes between perceptual and categorial content plus a prior awareness of the world, but it also qualifies as inadequate knowledge because its content lacks abstraction; and (3) essential intuition of abstract entities qualifies as fallible but adequate knowledge: adequate because it grasps the object's constant and necessary essence such as numbers, Plato's Forms, etc., and fallible because it is subject to further correction through more experiences. As we have already seen, this idea of correction is the only warranty for the epistemic value of intuition in many contemporary philosophers' views, such as Sosa's, BonJour's, and so on.

1.5.2 Epistemic Intuition in Contemporary Epistemology

As it was often said, Gettier's examples were the kind of entry that warranted the ongoing debate on intuition. For since then, philosophical discussion about the appeal to intuitions has become a familiar phenomenon in contemporary epistemological literature. Frequently, epistemologists start with some target claim for evaluation and then present one with a thought experiment that contains as part of a whole an intuition that has some bearing on the claim under evaluation. It is within the context of a thought experiment, then, that intuition gets its appearance. In (Dennett 1984)'s words, thought experiments are "intuition pumps." Intuition, therefore, lies at the heart of much current epistemological activity and the aim of

[29] Chudnoff's phenomenological view of intuition will be discussed again in more detail later in this book.

[30] It would take us too long to completely explain this Husserlian classification of intuition. A few words, however, should be said about it.

this section is to show how the phenomena of intuition gained great prominence in recent analytic epistemology by virtue of thought experiment.

Let us apply this idea to the examples we considered in Sect. 1.3.2. It was often argued that the above thought experiments are dependent on intuitions. For example, Jackson's thought experiment of Fred was widely considered by many contemporary philosophers to be relying on intuition; namely, our having all physical information about the subject won't help us to know everything about his color experience. That is, even after we know everything about his brain processes, physiology, history, body, etc., something still will be left out. Similarly, in his Robot Mary case, Jackson's intuition that lies behind this thought experiment is that the subject's previous physical knowledge is incomplete. That is why, after releasing her from her black and white room, she will acquire more than all the physical information already known to her.

Similarly, Searle's Chinese room case depends on the intuition that the computer lacks cognitive states. In the same way, Chalmers' Zombie case relies on the intuition that the realm of Zombies or Zombie itself (the subject of the story) is, indeed, conceivable. The same story goes for Cohen's Lottery cases. In case 1, the intuition is that the subject does not know that he will lose. In case 2&3, the intuition is that the subject knows that he will lose. Also, in Lehrer's Truetemp case, the author constructs his thought experiment on the basis of the following intuition: the subject, Mr. Truetemp, does not know what the temperature is when he truly says what it is. By the same token, Goldman's fake barn case involves two intuitions that are 'Henry knows' to 'Henry does not know.'

In Gettier's example we discussed, it is intuitive that Smith does not know the proposition 'the one who will get the job has ten coins in his pocket' is true. In fact, he was very lucky to believe it as true. In like manner, Kripke's Gödel–Schmidt thought experiment rests on the intuition that we do not refer to Schmidt when we use the name, Gödel. Finally, BonJour's example of Norman rests on an intuition that externalism violates: the subject is epistemically unjustified in believing that the President is in New York City.

This opens up the intriguing question of what intuition is.[31] Although there is no general consensus among contemporary philosophers about what intuition is, clusters of them reduce it to some other cognitive state such as belief (Lewis 1983), non-inferential belief (Gopnik and Schwitzgebel 1998; Devitt 2006, p. 491; Kornblith 1998), inclination or disposition to believe (Inwagen 1997, p. 309; Earlenbaugh and Molyneux 2009; Sosa 1998), sui generis mental state with propositional attitude namely, seeming (Bealer 1998a, 2002; Pust 2000; Huemer 2001, 2005a; Chudnoff 2011a; Bengson 2015; Koksvik 2011), inclination to judgment (Cohen 1986), and "spontaneous mental judgement" (Goldman & Pust 1998, p. 179). This list goes on and on.

[31] While chapter two will be entirely devoted to the analysis and development of an account of the nature of intuition, I will here lay out some of its prevailing definitions for purposes of clarity.

In summary, there is room for disagreement about the nature of intuition. However, a question arises as to whether this divergence can be divided into two types. One type lies in the variation of thought experiments' intuitions. And the other lies in the variety of what at least some intuition-theorists consider as intuition in their accounts. My principal concern in the next section is to emphasize this division by trying to explicitly identify what the epistemic consequences that follow from this action of considering such a divergence are.

1.5.2.1 Intuition in some Intuition-Theorists' Accounts

We just saw how different philosophers have different understandings of the term 'intuition.' But much is also left to be said. Some intuition-theorists, while arguing for their accounts of what intuition is, give examples of intuition that are different from the aforesaid thought experiment intuitions. This, with its consequences, will occupy the remainder of this chapter. In the next section, I will show that maybe it is improbable to have a general agreement on what intuition is. But it is unacceptable to claim that an account of intuition is conclusive, where in fact, it excludes thought experiment-generated intuitions. Whatever theory of intuition one has, one must be careful to keep in mind that it is thought experiments that generate intuitions. Here, I will begin by discussing what is considered as intuition by some contemporary intuition-theorists, outlining and discussing several examples from their accounts.

1.5.2.1.1 Chudnoff's Phenomenology-Based View

In his book, *Intuition*, Elijah Chudnoff tries to elaborate on and defend a phenomenology-based view of intuition, according to which "intuition is best understood as "a form of intellectual perception" (Chudnoff 2013a, p. 1). To that end, he establishes an alleged intuition-perception analogy: both intuition and perception are experiences with immediate capacity to justify beliefs making us aware of their subject matter. Yet, they are different in two senses. One is that they have two different subject matters. Another is that perception involves sensory experience, but intuition may and may not do so. To prove this view, he gives many examples. My aim here is not to argue for or against his view but simply to cite some of his examples of intuition. Doing so will help us to know what kind of intuitions he talks about and to what extent they are close to thought experiment intuitions. To save breath, ink, and energy, I will present his main examples within their contextual background in brief.

For exposition and defense of the idea that intuition has a distinct subject matter, he gives the following geometrical example: "you can intuit that all circles are symmetrical about all of their diameters" (Chudnoff 2013a, p. 10). Likewise, in order to put intuition experience in the scope of presentational phenomenology, he gives the following two geometrical examples: one is "two circles can have at most two common points." Another is "if a quadrilateral is inscribed in a circle, the sum of the products of the two pairs of opposite sides is equal to the product of the diagonals" (Chudnoff 2012a, pp. 58). His idea is that the former has presentational phenomenology, but the latter does not have this feature. However, Chudnoff also attempts to reply to the advocates of the claim that intuition has no presentational phenomenology. To that end, he differentiates between levels, so to speak, of intuition: the intuition that picks out intuition experience and intuition that picks out intuitive belief. The first has presentational phenomenology, but the second does not have this feature.

He first argues that in the case of perception, it is wrong to equate perceiving and to recall the sky as blue. And he moves on to say that the same story is true in the case of intuition by giving the following arithmetical example: "Suppose you intuit that 1 + 1 = 2, take your intuition at face value, and so form an intuitive belief that 1 + 1 = 2. Later, you might recall the intuitive belief that 1 + 1 = 2 ... it would be wrong to argue that some intuitions lack presentational phenomenology because this intuitive belief lacks presentational phenomenology. It would be wrong, because it would be committing the fallacy of equivocation" (Chudnoff 2013a, p. 59). That is, it would be confusing between the aforesaid two levels of intuition.

Having established the epistemology of intuitions on the basis of their phenomenology, Chudnoff goes on to explain what are the ingredients of intuition as experience consists of by giving the following arithmetical example: "the bigger of two numbers is the average of their sum and difference—max (m, n) = [(m + n) + |m – n|]/2" (Chudnoff 2013a, p. 63). For him, at first glance, one will be neutral regarding the true value of this intuition. After a while, one will get to know that this is the case. What happens in between is reflection along with other mental states such as conscious thoughts, imaginings, and so on. At any rate, we will see below that Sosa offers an alternative view that opposes two of Chudnoff's major claims: intuition as experience and the analogy between perception and intuition.

1.5.2.1.2 Sosa's Competence-Based View

In "Minimal Intuition," Ernest Sosa attempts to provide an understanding-based view of intuition, according to which intuition is belief about abstract propositions gained by understanding only, without any role played by any other source of knowledge such as reasoning, perception, inference, and so on. Sosa gives us the following examples of it: "that 2 + 2 = 4; that no cube is a sphere; that nothing is numerically self-diverse" (Sosa 1998, p. 262). It seems that he thinks that to anyone reasonably competent in basic arithmetic, it will be obviously true that 2 + 2 = 4.

He then moves on to consider two main objections to this view of intuition. The objections under consideration are these: it introduces intuition being thin and unreliable. His response to the first objection is that, in his view, intuition may be fallible, but its fallibility is due to "our performance" of "our reasoning competence" rather than the competence itself. To respond to the second objection, he emphasizes a comparison between intuition and other sources of knowledge and argues that intuition is reliable, at least in normal circumstances, and generally appropriate for exercising the competence of intuition.

Later on, in a series of publications, Sosa developed this idea of competence or virtue model of intuition. He first rejected what he calls the perceptual model of intuition, according to which intuition is analogous to perception. His pressing rejection comes from the base that in perception, sensory experience mediates between belief and its object. While in intuition, no experience plays any mediative role as such. Otherwise, the truth of intuition will be an episodic phenomenon. Therefore, the perceptual model is not appropriate for intuition. In addition, he gave up what he calls the introspective model of intuition, which can be considered to be justifying fallacious reasoning, as in affirming the consequent in the following false intuition: "that, necessarily, if q, and p→ q, then p" (Sosa 2007a, p. 58). Consequently, the introspective model is, he thinks, also not appropriate for intuition. Sosa, accordingly, introduces his competence model according to which intuition is prima facie true when it is so if it manifests an epistemic competence (or virtue). And when it is false, it will be so because of an error related to the subject's performance of her competence and not to the competence itself.

1.5.2.1.3 Bealer's Sui Generis Mental State-Based View

Begin reading on the issue of intuition in contemporary analytic philosophy; it will not be long before you stumble upon George Bealer's own view of the very issue. Bealer has provided, in several writings, a phenomenological account of intuition, according to which the philosophical method or what he calls "Standard Justificatory Procedure" (or the procedure of a priori justification) relies on rational and a priori type of intuition. Arguing that intuition enables philosophy to provide independent and stronger answers to the common questions between science and philosophy than science can do, he holds that the thesis of the autonomy and authority of philosophy with respect to empirical sciences is true, unlike how scientific essentialists habitually used to think.

On the present view, intuition is an intellectual (not experiential) seeming or a *sui generis* mental state with an episodic character. A state as such qualifies, though it is fallible, as a reliable basic source of evidence (or as the reason) to justify our theories and beliefs in virtue of its modal tie to the truth. It has a tie as such because it consists of semantically stable concepts that an individual possesses a priori and understands determinately, not nominally. However, to understand the Bealer-style account of intuition, one has to properly understand a matrix of other expressions such as rephrasal strategy, conceivability, epistemic possibility, metaphysical possibility, modal error, etc.[32]

[32](Bealer 1996a) adopts the Kripkean rephrasal strategy to correct the errant modal intuitions. Kripke's strategy was "to handle the apparent contingency of certain cases of the necessary a posteriori" (Kripke 1980, p. 150). I will not pursue it further here, especially that Bealer; later on, presents many arguments to reject it in (Bealer 2004).

As we have seen, philosophy, for Bealer, is committed to intuition, which is, in turn, committed to modality. He, therefore, uses the modal character to dye both intuition and philosophy equally. A few examples will suffice to make us know the kind of intuition he regularly talks about in his writings. Here are the examples: "if P then not not P" (Bealer 1998a, pp. 205, 209, 211, 217; 1996a, p. 127; 1996b, pp. 5, 11); "it is impossible that P iff it is necessary that not P" (Bealer 1998a, p. 201); "De Morgan's laws" (Bealer 1996a, p. 123 & Bealer 1998a, p. 207 & Bealer 2008, p. 190 & Bealer 1996b, p. 5); "the naive comprehension axiom of set theory" (Bealer 1996a, p. 123 & Bealer 1996b, p. 6 & Bealer 1998a, pp. 202, 208, 209 & Bealer 2008, p. 190); "Mathematical limits" (Bealer 1998a, p. 211); "intuitions about simultaneity and Euclidean geometry" (Bealer 1996a, p. 123 & Bealer 1996b, p. 9); and so on and so forth.

With regard to the three intuition-theorists' examples of intuition, it is clear that they are logical and, to a greater extent, mathematical intuitions. One can object that this particular sort of intuition may have nothing to do with philosophical intuitions in general and epistemic ones in particular unless Aristotle, for example, was wrong and, accordingly, all branches of knowledge are expected to have the same degree of mathematical precision. Furthermore, let us leave this point for the sake of argument and ask: is this sort of intuition the same one, which was used in the thought experiments of analytic epistemology?

Taking a look at the above-discussed ten cases, we can comfortably say no; they are not the intuitions involved in the standard intuition-based philosophical practice of thought experiments in contemporary analytic epistemology. At any rate, in the sense of the target object involved, there is an analogy between mathematical and thought experiments' intuitions. Both involve abstract objects. But in the sense of content, the analogy is wrong. For while the content of mathematical intuitions is mathematical truths, the content of epistemic intuitions is a type of mental state.[33] However, some might say that maybe without this kind of mixing up these two different things (i.e., mathematical and thought experiment intuitions), those theorists would not be able to establish their accounts of intuition. That may be true, but, as I will show in the next part, doing this reduces intuitions to the world of Platonic forms.

1.6 Mathematical Platonism

I will not be arguing that the examples of intuition given by a particular position of the three intuition-theorists and thought experiments' intuitions are the same. Nor will I argue for or against the claim that the examples of intuition given by any of the other two positions just mentioned and thought experiments' intuitions are not

[33] There may be many other ways to argue that this analogy does not hold. But discussing these plausible alternatives is out of my working framework. I simply wanted to differentiate between mathematical and thought experiment intuitions, and what was mentioned is enough for that purpose.

different. At the end of the previous section, it was firmly resolved that they are dissimilar. Accordingly, my point in this section is largely consequent: suppose we have reason to believe that the examples of intuition given by these three intuition-theorists depict the sort of intuitions used to arise in the thought experiments of contemporary analytic epistemology; what is the thing that follows from a supposition as such.

According to Plato, true knowledge enjoys the following three features: it must be certain, unchanging, and universally acceptable. What meets those three conditions is the knowledge of abstract mathematical objects. Mathematics is considered to be a stepping stone to the quest for true knowledge. Mathematics deals with concepts or forms (like triangularity) that are objects of true knowledge (a triangular block). Now, if those three intuition-theorists hold that intuitions are mathematical entities, it follows that they look at intuition as capturing Platonic forms. In such a circumstance, as (Goldman 2007) holds, intuition captures some sort of eternal, non-spatially located entity that exists entirely outside the mind, like the form of the Good, for example. The thesis under consideration, then, is this: it seems true to believe that these intuition-theorists are committed to mathematical Platonism.

Perhaps, such a view is in need of motivation in light of the following reflection. Upon considerations, it would not be fair not to indicate that those intuition-theorists do sometimes discuss intuitions that emerge from contemporary epistemic thought experiments. For example, (Bealer 1998a) discusses Jackson's Mary case (p. 205) to show the dominance of the philosophers' use of intuitions while philosophizing. He also discusses Burge's arthritis case (pp. 205, 208, 221) as an inappropriate example of intuition; it misleads us in achieving determinate possession of semantically stable concepts, the necessary condition for the reliability of intuition. Bealer goes beyond that and claims that thought experiments involve only physical intuitions (Bealer 2002, p. 74).[34]

In the same way, (Chudnoff 2013a) discusses Goldman's fake-barns case (pp.181–3, 192–3) and Gettier's cases (pp. 77–8, 142, 181, 192–3, 200). But these cases were discussed in order to draw some similarities between thought experiment intuitions and geometrical intuitions. These analogies were claimed in order to say that "philosophical intuitions are phenomenally like logical and mathematical intuitions, just different in their subject matter" (p. 76). Thus, his phenomenological view of mathematical intuitions also goes for philosophical intuitions, he claims. That is why his book is full of examples of mathematical intuitions. However, one may object that if this analogy is true, the argument could go in the opposite direction. That is, he could make his book full of examples of philosophical intuitions claiming that they are phenomenologically analogous to mathematical intuitions. What about Sosa? He does not discuss thought experiment intuitions even, at least

[34] Bealer claims that there is an important distinction between intellectual seemings or a priori intuitions that present themselves as necessary and physical intuitions elicited by thought experiments that present themselves as contingent. While philosophical inquiry typically relies upon the former, empirical sciences rely upon the latter.

not to my knowledge of his works. This means that Sosa's favorite examples of intuition are restricted to mathematical and logical truths.

What all this shows is that by intuition, they understand mathematical intuition over thought experiments' intuitions. But even so, there are several problems with this view. For instance, one may object that the more mathematical the intuition, the harder it becomes to account for without being loyal to Platonism. To simplify, the above three criteria of knowledge can be used to find out if something is true knowledge or not. We use these criteria to eliminate different sources of knowledge that do not fulfill these criteria, like perception and memory, for example. For Plato, true knowledge is knowledge of virtues. There are three types of virtues: Truth, Beauty, and Goodness. The highest virtue, according to Plato, is Goodness. For Goodness is not a quality of Truth and Beauty. It is their essence. Truth is good and Beauty is good. Why is the knowledge of virtues true knowledge? He believes that true knowledge must bring about betterment or change in the character of a person. If you know what right is, then you won't do what is not right. If one does not do so, then the knowledge of this value is incomplete. And untrue knowledge or ignorance happens when one's virtue of knowledge is incomplete.

In the same way, Sosa proposes an intellectual virtue of intuition as a warranted source of true knowledge. While this epistemic virtue is indeed reliable, the subject's errors in performing its mechanisms make it produce false intuitions. Therefore, I propose that Sosa's intuition–virtue qualifies as a virtuous source of knowledge according to the criteria of Plato's realm of Forms. Moreover, Bealer holds that the question of intuition or any central philosophical question has some features such as universality (noncontextuality), generality (not pertaining to a certain individual, species, or historical event), etc. (Bealer 1998a, pp. 203–4). These correspond respectively to Plato's universal acceptability and unchangeability as two out of the three conditions true knowledge must meet. In addition, (Chudnoff 2014a) adopts Gödel's perceptualist view of mathematical intuitions. And the most distinctive aspect of Kurt Gödel's thought is certainly Platonism about mathematical objects and knowledge (Parsons 1995). As a result, those three philosophers did not get rid of the belief in a separate world of abstract mathematical entities, and then they fell victim to Platonism about intuitions.

Viewed in this way, intuitions become copied instances of a Form of intuition that causes them. According to Plato's theory of Forms, Forms are ideal or abstract, and particular entities in the physical reality are copies of them. For instance, Beauty is a virtue, and one's actions, those that count as beautiful, are instances of it. If this line of thought is true, this amounts to the view that instances of intuition are derived from INTUITION as a Platonic Form. The danger of this view is that Plato considers Forms as independent from the human mind. That is to say that if intuition is a Form, intuition is no more created by the mind. Rather, an instance of intuition turns out to be determined by a Form of intuition to which it must also conform. Here those philosophers create a dilemma; they are committed to Platonism, according to which Forms are independent of the human mind, and at the same time claim that intuition is dependent on the human mind. Whether intuition is an intellectual perception, a belief about an abstract proposition gained by understanding, or a sui generis mental state of seeming, it is still dependent on the human mind.

On the other hand, suppose that those intuition-theorists are right: a particular intuition is derived from INTUITION as a Platonic Form independently from the intuiter's mind. This raises a difficult issue concerning the question of what connection is there between the particular intuition and its Form. Take any of their mathematical examples. Or, equally, take, for example, BonJour's intuition that Norman is not epistemically justified in believing that the President is in New York City. As noted above, we must recognize that the intuiter of any of their mathematical examples or even BonJour acquired the relevant intuition by passively receiving it from its Form. A further and much more interesting consequence of Platonism is that intuition fails to be a mental state. For if intuition is independent of the human mind, then it cannot be a mental state. That is, if Platonism is true, then intuition is not an intellectual perception, a belief about an abstract proposition gained by understanding, or a sui generis mental state of seeming.

Either Platonism about intuition is true. Thus, intuition fails to be a mental state of any kind. Or Platonism about intuition is false. Therefore, these intuition-theorists should stop giving examples of mathematical intuition as the finest examples to illustrate or exemplify what intuition is, especially when the debate is on thought experiment-generated intuitions. Either way, they are wrong.

In the next section, I will explain and defend an anti-Platonist approach to the issue of intuition.

1.7 Epistemology without Intuition[35]

It is common in the philosophy of mathematics to discuss mathematical questions related to epistemology. Whether it is with reference to Pythagoras, or Plato, or Aristotle, or Leibniz, or in the more recent times with philosophers like Frege, Russell, etc., epistemological issues have always occupied a special place in some form or the other in the philosophy of mathematics. Wittgenstein is also famous for his formulations in the philosophy of mathematics.

What follows will be an overview of what I call Wittgenstein's anti-Platonic argument. Wittgenstein appears to think that intuition is the situation in which one "knows immediately which others only know after long experience or after calculation" (Wittgenstein 1976, p. 30). The preceding claim, then, is that intuition is a spontaneous mental activity that occurs without reasoning. An initial consideration in favor of intuition is that, according to him, mathematics needs intuition. The source of intuition is language. Language is a human activity of non-experiential calculating procedure (Wittgenstein 1922, p. 83). If Wittgenstein is right, then mathematical intuitions are not derived from the world of Forms, and accordingly,

[35] This part is mainly derived from my paper (Hamdo 2018), unless otherwise indicated.

Platonism is false. I shall argue that this is true, but Wittgenstein's argument ends up with a different problem, namely, an epistemology without an epistemic role for intuition.

Now, upon reflection, the description of intuition I just mention does not only mean that the process of calculation is not based on empirical intuition, but it also means that the very process provides the a priori form of intuition representing the experiential world. I said 'representing' and not 'describing' because, for him, mathematics does not describe things. Instead, it is the way we describe things (Wittgenstein 1922, p. 85). Traditionally, philosophers used to establish the origin and possibility of knowing necessary truths on a firm or permanent basis of the role they play. This story does not go for Wittgenstein, whose aim is to know by virtue of what a certain proposition is necessarily true. His view is that the truth of a proposition depends on our usage of it and not on the role it plays (p. 82). So, unlike Platonism, the view that mathematical proposition is concerned with a distinct world of truths that are abstract entities, Wittgenstein asserts that mathematical propositions refer neither to abstract nor to empirical entities. Nor are they accountable by the law of excluded middle.

Having regarded language as a human activity of a non-experiential calculation process, he argues that a mathematical proposition is a linguistic statement which is a rule for our usage of signs (Wittgenstein 1978, p. 243). This leads us to what he calls a sign-game, according to which mathematical propositions may be applicable outside mathematics, but this does not mean that mathematics must empirically be applicable. For him, mathematics is both pure and applied (Wittgenstein 1978, pp. 257, 232, 258–60, 295, 376). It is pure or non-applied when it is not tied to empirical applications, as in establishing that 16 x 14 = 224. It is applied when one makes a mathematical statement meaningful by empirically using it in or applying it to reality, as in applying any arithmetic theorem. However, the applied aspect of mathematics makes it a part of the history of human beings (Wittgenstein 1978, pp. 92–3, 176,182, 399).

Moreover, it can also be interpreted as an attempt to argue against a psychologistic approach to intuition. That is why Wittgenstein holds that intuition is free from any psychologistic aspect (Wittgenstein 1958, §§ 186 & 187). He maintains the idea that following a rule of succession does not rest on a prior mental grasp of the succession. Therefore, such a mental grasp does not explain rule-following, or it has no role to play in knowing how to follow a rule. To emphasize this point, he introduces a new word, 'technique' (Wittgenstein 1958, § 199). By the term 'technique,' he means, in particular, the technique of counting. He writes: "there is no discovery that 13 follows 12. That's our technique- we fix, we teach our technique that way" (Wittgenstein 1976, p. 83). His key thought is that what justifies the series of natural numbers is not a certain mental grasp. What does so is the fact that they are already given in the custom or technique of counting? Unlike the Platonist picture of ideal abstract objects, Wittgenstein defends a type of reasoning related to rules or norms of representing things. For example, a mathematical proposition is a priori because it is an intelligible description of reality (Wittgenstein 1978, pp. 363, 425, 431). It is normative because it is the norm of applicable representation; it can be applied to a pyramid (Wittgenstein 1958, § 139). Now, when a rule is correctly followed, in virtue of what does it get obeyed by people?

Wittgenstein's response to such a question is that a rule gets followed because technique produces consensus. Otherwise, it won't be called a rule. In (Wittgenstein 1958, § 142), he holds that people won't correctly apply a rule that looks abnormal to them. Here it may be argued that technique cannot be a matter of stipulated convention because mathematical theorems cannot be true just because of their correspondence to conventional wisdom. In my reading, he adopted moderate conventionalism, according to which technique is customary of rules/norms and a way to apply them by the community. If you can imagine mathematics before it got axiomatized by the Greeks, you can understand what he meant by technique (Wittgenstein 1978, pp. 232—4).

As far as I can tell, if Wittgenstein is right, then it follows that Platonism is false. The reason to entail the denial of Platonism is owed to the fact that Wittgenstein asserts that mathematical proposition is not part of the alleged world of the truths of abstract entities. Instead, it is a linguistic statement whose truth value has certain conventionality to it. In such a circumstance, intuition is no more considered as some sort of eternal, non-spatially located entity. In my view, for him, intuition is part of human mathematical activity. In his words, "what interests me is not having immediate insight into a truth, but the phenomenon of immediate insight. Not indeed as a special mental phenomenon, but as one of human action" (Wittgenstein 1978, pp. 32). One can say that Wittgenstein gives us an alternative idea of how one can intuit a mathematical truth without appealing to platonic entities or even to infallible intuitions.

Suppose that those intuition-theorists' accounts of intuition do not suffer from a contradiction arising from being too Platonic and consider intuition to be a mind-dependent mental state, whether as an intellectual perception, a belief about an abstract proposition gained by understanding, or a sui generis mental state of seeming. If Wittgenstein is right, it is not clear whether mathematical intuition can still be best understood as a mental state in any of those intuition-theorists' sense of the word. Actually, perceiving the grasp of rules as a community driven normative exercise makes it exist entirely outside the mind. This entails that if there is any epistemic role to be granted to intuition, it must be attributed to the norm governing it and not to intuition as a mental state. If so, then although Wittgenstein may successfully avoid Platonism, his view is subjected to the problem of not considering intuition as a mental state. As a result, we have come to the conclusion that if Wittgenstein is right, then from now onwards, intuition as a mental state should disappear from epistemology, and accordingly, it no more plays any epistemic or evidential role. That may well have been something Wittgenstein would have applauded.

However, considering the fact that thought experiments generate intuitions and these are used in philosophy extensively, we need an account of intuitions such that we know what their evidential status is and what their nature is. We know that too much reliance on mathematical examples may not do much to answer either question. Wittgenstein's views may cause Platonism to retreat but leaves us wondering about the current status of the use and evidential value of intuitions. We need to investigate further to arrive at any strong conclusions.

1.8 Concluding Remarks

In this chapter, I have shed light on how thought experiments are used by scholars in a diverse set of fields. Subsequently, I have illustrated the emergence of thought experiments, particularly in philosophy. I have shown the interdependence between thought experiment and the history of philosophy (western and eastern) from pre Socratic philosophy to the present. Then, in order to put the subject matter of the chapter in the right track, I have narrowed down the scope of focus by making it limited to epistemology in particular. Discussing some representative thought experiments, both from classical and contemporary epistemology, I have attempted to characterize epistemic thought experiment along with trying to explore its structure and function.

As a result, I have developed my own definition of thought experiment, showing how intuition is an inevitable constituent of thought experiments in contemporary analytic epistemology. Since it was often argued that thought experiment really does what it seems to be doing in virtue of its reliance on intuition, my concern became the nature of intuition. To that end, I have reviewed some examples of intuition given by three of the most distinct intuition-theorists' accounts showing how their examples of intuition are different from thought experiments' intuitions. After this, I have argued that these intuition-theorists are committed to Platonism. And, I have shown how Wittgenstein got rid of Platonism but left epistemology without intuition as a mental state. Accordingly, the epistemological concern remains the same: what is the nature of intuition? If philosophers want to do epistemology and do it well, then we need an answer to this question. The next chapter is devoted to the task of answering this question.

Chapter 2
The Nature of Epistemic Intuition

2.1 Introduction

In the last chapter, we saw that epistemic intuition is an essential constituent of epistemic thought experiments. The question then arises: what is the nature of epistemic intuition? To answer this question, I reviewed some examples of intuition given by three of the most distinct intuition-theorists' accounts. I argued that these intuitions belong to a Platonic kind of realm and, accordingly, are different from intuitions pumped by epistemic thought experiments. As a result, the question remains the same: what is the nature of thought experiment-generated intuition? It is now time to attempt to develop one core concept of what epistemic intuition is so that we can assess the merits of the various types of epistemic theorizing. It is my contention that there is a better form of epistemic intuition than the Platonic one. This conception of epistemic intuition is important in that it will serve to determine whether the various types of analysis concerning the structure of epistemic intuition are plausible in yielding our epistemic goals (whether they are justified by a good or legitimate reason). To this end, the goal of this chapter is devoted to developing such a conception.

In other words, I am going to develop an account of intuition that captures the appropriate form of intuition for the appropriate form of epistemological analysis. This account will go beyond the brief explication of intuition given at the end of chapter one. This chapter will be organized in terms of the possible features of intuitions, not in terms of people who have provided accounts. Consequently, I will initially present and discuss substantive accounts of intuitions when they have the feature that I am discussing in that section. I do not intend to attack or defend anyone's account directly per se; rather, I will attack or defend the necessity of particular desiderata. The rejection or defense of anyone's account is purely tangential to my overall goal.

M. Hamdo, *Epistemic Thought Experiments and Intuitions*, Philosophical Studies Series 150, https://doi.org/10.1007/978-3-031-33480-1_2

I am going to develop an account of the nature of intuition according to which epistemic intuition is not captured by the traditional distinction between a priori and a posteriori way of knowing. Instead, intuition involves several dimensions or aspects: a priori, a posteriori, both a priori and a posteriori, and neither a priori nor a posteriori. Moreover, keeping in mind that it is directed to an object, I will be arguing that epistemic intuition is an intentional but non-propositional mental state that has no special phenomenology. In addition, I shall be trying to show that epistemic intuition can be individuated by virtue of its home, thought experiment. Here is the plan for the chapter.

The chapter is divided into five parts. In the second part, I will attempt to discuss three epistemic discourses given by recent epistemologists who identify intuition as a mental state. In the third part, I will be considering three ascriptions attributed to intuition by these three discourses: intuition is a priori, propositional, and a phenomenal mental state. Contrary to that, I shall be arguing that intuition transcends the a priori-a posteriori distinction. Putting it another way, I will be discussing each of those aspects, showing when intuition possesses one of them, when it has none, and when it has all of them. Subsequently, I will try to prove that mental states, including intuition, are intentional though non-propositional. Next, I will attempt to show that intuition has no distinct phenomenology. In the fourth part, I will demonstrate that intuition is individuated by the particular place where it arises, thought experiment. I will devote the fifth part to sketching some concluding remarks of the chapter. So, in the end, my picture of the nature of intuition will be different from what is usually assumed by many who concern themselves with the topic of what intuition is.

2.2 Intuition as Mental State

The issue of intuition arouses considerable interest when it is raised in epistemological analysis. Furthermore, each analysis made different claims about the type of intuition at hand. To many, it is obvious that intuition must be considered a linguistic disposition. For example, Richard Miller writes that "intuition is … linguistic disposition as manifested in the imagined circumstances" (Miller 2000, p. 235). This is to be contrasted with a trend by some philosophers who treat intuition as a mental state. By 'mental state,' it meant the state of mind the subject is in while intuiting. In what follows, I will be discussing three forms of such a trend. In the previous chapter, I have looked only at some examples of intuition given by these intuition-theorists' accounts. Let us, in what follows, look at their accounts of the nature of intuition in the detail they truly deserve. So, in this section, I will look at the aspects of intuition according to these three accounts given by present epistemologists refining some and rejecting others. However, while presenting objections to them, I will be paving the way to develop my own conception of the nature of intuition.

2.2.1 *Bealer's Intellectual Seeming Analysis*

George Bealer has, in a series of papers, come up with a comprehensive and perspicuously defended account of intuition that describes the distinctive features of intuition. The cornerstone of Bealer's account of intuition is his understanding of intuition in phenomenological terms. That is, intuition is a form of phenomenal seeming or appearing. In this sense, intuition is just what seems true to the subject or what she feels to be true. He writes: "for you to have an intuition A is just for it to seem to you that A" (Bealer 1998b, p. 271). So, it would seem obvious to Bealer that intuitions ought to be treated as a mental state. But this requirement is not just that intuition is a mental state, but a certain mental state. For a better understanding, it might be helpful to view his account as being divided into two arguments.

The first is a negative argument, in which he tries to tell us what intuition is NOT, attempting to refute others' way of characterizing intuition. By providing a cluster of negative characterizations of intuition, in fact, he refuses to confuse intuition with other mental states. For instance, one should not mix intuition up with belief because "belief is not a seeming; intuition is" (Bealer 1998a, p. 208). I may believe what I do not have any intuition about, as in many mathematical theorems. Contrariwise, I may intuit what I do not believe. I may intuit that set theory is true but believe it is paradoxical. The same story goes for the Muller-Lyer illusion, where I may intuit that the lines are not of the same length but believe that they are congruent.[1] In both cases, intuition and perception, seeming, is present regardless of the presence of belief. To lend more plausibility to this view, he points out that "belief is highly plastic; not so for intuition" (Bealer 1998a, p. 208). One often doubts one's beliefs but rarely disputes one's intuition. Furthermore, intuition cannot be confused with a spontaneous inclination to believe (Bealer 1998a, 1996c). I may have an attraction to believe in eternal life without having any intuition about eternality. While intuiting, some cognitive processes must happen episodically to the subject, but the inclination to belief is not episodic. I may be spontaneously inclined to believe De Morgan's laws as false, but after reflection, I start finding them true. If intuition is an inclination to belief, these laws would seem to me spontaneously (not episodically) true. Another proposal equates between intuition and conscious belief raised from unconscious background beliefs. Although Bealer does not rule out the notion that non-conscious mechanisms may play some role in formulating intuition, he

[1] Here, he draws a kind of intuition-perception analogy. He gives the optical illusion of Muller-Lyer as an example of intuition that must be differentiated from belief. It seems that he tries to say that intuition should be acceptable as seeming just like perception. The case is of interest because intuition-perception analogy, as we will see later in this book, is not a part of Bealer's conception of intuition.

At any rate, the Muller-Lyer illusion is an optical illusion devised by Franz Carl Muller-Lyer. It consists of three arrows of equal length. When subject to it, it still seems to one that one arrow is longer than the other, even after one has convinced oneself otherwise by measuring the lines. What explains the Muller-Lyer illusion is the fact that the lines are processed by the visual system as a three-dimensional scene. This is the reason why one gets to see one arrow as longer than the other.

rejects this proposal. For our non-conscious background beliefs about a topic could be opposed to one another. This contradiction prevents us from having an intuition about the very topic. I may have a non-conscious background belief about something without having an intuition about it. For example, I may have a non-conscious background belief that I was not born in India without having any intuition about that. Also, if this proposal (i.e., the proposal of non-conscious background intuition) is true, then intuition becomes just a copy of the unconscious background belief one already had (Bealer 1998a). We can conclude that the proposed account does not do justice to the matter of the novelty of intuition.

What is more, intuition is different from sense perception, judgment, guessing, hunch, memory, common sense, linguistic intuition, conceptual intuition, and report of consistency. Intuition is not sense perception because the former is intellectual and the latter is perceptual (Bealer 1987a, 1998a). What seems to you intellectually the case may not seem to you sensorily so and vice versa. For instance, mathematical truths do not seem sensorily the case, and the number of cells in your body does not seem intellectually the case.[2] Moreover, "judgments are a kind of occurrent belief; as such, they are not seemings. Guesses are phenomenologically rather more like choices; they are plainly not seemings. And hunches are akin to merely caused, ungrounded convictions or non-inferential beliefs; they too are not seemings" (Bealer 1998a). Similarly, recalling something from your memory is phenomenologically different from intuiting it. In memory, the feeling of 'pastness' does present, but in intuition, it does not. When you remember that $18 + 17 = 35$, you feel that you are recollecting something you have already learned. While it just seems to you, when confronted with, that $2 + 2 = 4$. When you remember a theory you have already proven, it may not just seem to you; it is the case.

Also, intuiting that P is different from P's being just true by common sense. For common sense can be identified with elementary intuition only. But since not every intuition is elementary, so intuition is not common sense. It is also wrong to claim that all intuitions are linguistic ones, especially if linguistic intuition is understood in terms of words of a certain language and their applicability. I, as a non-Hindi speaker, can have an intuition in Hindi though I do not have intuition about Hindi words and their applicability. Intuition then does not concern words of a certain language. It concerns concepts, where 'concept' is understood in a broad sense and not as a cognate word of 'analytic' simply because not all intuitions are analytic.[3] The same story goes for considering intuition to be the same as a report of consistency (Bealer 1998a). The main thing that differentiates intuition from the aforesaid mental states is that intuition is phenomenologically different, i.e., its being

[2] A clearer explanation of the point of the previous footnote is provided here. The analogy between intuition and perception is acceptable only in the sense that, in both cases, seeming is present. We should not go beyond that confusing intuition with perception. Instead, as we will see later in this book, Bealer's conception of rational/a priori intuition excludes any perceptual element.

[3] If I understood the intended meaning correctly here, what Bealer means is that not all intuitions have intrinsic conceptual connections. Nor do we grasp the true knowledge of them by virtue of meaning only.

seeming, but other mental states are not. For example, S intuits P means that it seems to S that P. Perhaps this point can be elaborated more in the second argument.

The second is a positive argument in which he tries to tell us what intuition IS. On the positive side, not much is said beyond claiming that "intuition is an *intellectual* seeming," where seeming is understood as a unique, conscious, and cognitively reflective episode (Bealer 1998a, p. 208, 1992, P. 102).[4] Bealer maintains that a genuine "intuition ... is a *sui generis*, irreducible, natural ... propositional attitude which occurs episodically" ((Bealer 1998a, p. 213, 1996b, p 29, 1996c, p. 169). Consider, for example, Gettier's Smith-Jones case. When confronted with Gettier's intuition that 'Smith does not know,' it will first seem to you neither true nor false. But after a few moments, a conscious and cognitively reflective episode happens and makes you see the intuition to be the case. However, it is worth noting that by 'intuition,' he means 'a priori intuition' (Bealer 1996b, p 30, 1992, P 102). He draws a distinction between this type of intuition which he calls "rational intuition," and what he calls "physical intuition." Rational intuition expresses itself as necessary in the sense that it could not be otherwise. For example, "necessarily, if x intuits that P, it seems to x that P and also that necessarily P" (Bealer 1998a, p. 207). If so, if rational intuition is a priori and necessary, then by 'intuition,' he understands analytic intuition in the traditional sense of the term 'analytic,' though he, as mentioned earlier, rejects to identify intuition as a cognate word of analytic.

Physical intuition, on the other hand, does not strike us as necessary. Instead, it expresses itself as contingent in the sense that it could be otherwise, as in physical and natural laws. More importantly, "rational intuition is silent about this sort of question. Rational intuitions concern such matters as whether a case is possible (logically or metaphysically), and about whether a concept applies to such case" (Bealer 1998a, p. 207). Applied to Gettier's Smith-Jones case, we have two rational intuitions. One is that the case is possible. Another is that the concept of knowledge is not applicable to the subject in the example.

Putting aside the details of this characterization, it exemplifies a general strategy: S has an a priori and rational intuition that p, if and only if it intellectually seems to S that necessarily p. This way of understanding intuition makes the necessary part of the content of the intuited proposition. What he says is that every a priori intuition is necessary. This might be said to be imprecise. Let us begin with an obvious observation: the problem is the fact that under some circumstances, one's a priori and rational intuition may lack the property of being necessarily the case. There are situations in which one's a priori and rational intuition seem to be precisely the case, but not necessarily so. These remarks result in one of two conclusions. Either Bealer intends to offer this analysis solely as an account of intuitions in logic or mathematics. Or this analysis is applicable to the kinds of epistemic intuitions invoked in thought experiments. To evaluate the status of these conclusions, let us consider each of them individually.

---

[4] I have to admit I am not quite sure what 'a unique, conscious, and cognitively reflective episode' is. Put it into Bealer's context; I guess it signifies a mental event that takes place when it seems to someone that such-and-such is the case.

To begin, let us consider the status of the second conclusion. Take, for example, the Truetemp case. Let us say that Lehrer a priori and rationally intuited that the subject Mr. Truetemp, does not know what the temperature is when he truly says what it is because it intellectually seems to Lehrer that this is necessarily the case. But the case can be otherwise because "examining empirical studies of actual subjects who, like Truetemp, have received new perceptual faculties ... show[s] that Truetemp must have been endowed with all of the reorganized neural circuitry and cognitive skills that subjects with new perceptual faculties normally acquire during a long and difficult process of adaptation and development" (Beebe 2004, p. 307). If these studies are right, then Lehrer's a priori and rational intuition might occur during such a long and difficult process of adaptation and development and accordingly lack the property of being necessarily the case.

Suppose that Bealer does intend to offer his analysis solely as an account of intuitions in logic or mathematics. Even in such a case, the necessity of one's a priori, rational and intellectual seeming won't be warranted. Suppose, for example, that I a priori and rationally intuit that I can always connect two points with a straight line simply because it intellectually seems to me that it is necessarily the case. That will be the case so far as we consider it by Euclidean geometries that understand space as a plane and two-dimensional. But under some circumstances, the case won't remain necessary at all. It won't be the case so far as we consider it by non-Euclidean geometry that understands space as curved and not flat. This is not the only example in the town. Since the advent of mathematics to the present, mathematicians have frequently debated over the apparent necessity of their propositions that are equally intuitions but jointly incoherent. Whether it is with reference to non-Euclidean geometry, Cantor's theory of the hierarchy of infinite sets, or the recent theory of the continuum of elements that are noticeably different from each other, mathematicians have repeatedly debated over the necessity of their most potent intuitive convictions. On the strength of these sorts of examples, it seems that they provide the database for the formation and justification of the conclusion that necessity is not necessary for intuition.[5] However, some oppose this conclusion defending Bealer's position regarding necessity. For example, although he provides a different phenomenological approach to intuition, Sosa agrees with Bealer in many significant respects.

2.2.2 *Sosa's Understanding Analysis*

Bealer's considered account was shown to be unfavorable by Ernest Sosa, who, like Bealer, accepts that a person cannot account for intuition as a form of belief. Nonetheless, he differentiates between two senses of the intuition-belief distinction: the weaker sense, according to which not every belief is necessarily intuition, and a

[5] I will leave further consideration of this proposal until Sect. 2.3.

stronger sense, according to which being an intuition cannot be equivalent to being a belief. On Sosa's view, Bealer has given many arguments in favor of the intuition-belief distinction, but only one of these arguments can be classified under the stronger sense of the intuition-belief distinction. And this argument is the one that says that intuition and belief are significantly two different things. So, if the intuition-belief distinction is to be understood both in the weaker and stronger senses, then other arguments given by Bealer are not adequate to meet this purpose since they are classified under the weaker sense that does not firmly ensure that intuition and belief are substantially different (Sosa 1996). Nonetheless, according to Sosa, we can think better of intuition. Here is how.

Sosa has offered an account of intuition that attempts, unlike Bealer's account just offered, to deal with a reducible notion of intuition: intuition as intellectual seeming which is reducible to an attraction to assent a propositional content. Here is how he develops his argument. To Sosa's mind, Bealer asks us to think of intuition as intellectual seeming, which is similar to sensory seeming in visual perception. This entails that experience and seeming are analogous.[6] Contrary to that, Sosa argues that seeming is constitutively and functionally different from experience. They are constitutively different because experience does not need a rational underlying foundation. On the other hand, seeming does need a rational underlying foundation, namely, one's rational agency that plays the role of motivating, stative reason, or reason *for which*. In his words, "the seeming manifests the subject's rational agency; the experience does not." To clarify, he gives the example of the visual sensory experience of a Spanish word. In a fluent Spanish speaker, such an experience prompts seeming, but in a person ignorant of Spanish, it does not (Sosa 2014, pp. 36–39). In addition, seeming is conceptual; experience is not. One cannot intuit content that one cannot completely represent because of lacking the required concepts, but one can experience content that one cannot completely represent because of lacking the required concepts (Sosa 2013, p. 188; 2011, pp. 456–457). At any rate, experience is not only constitutively but also functionally different from seeming.

Suppose you see a table in your room. Here we have to distinguish between three things: your perceptual belief that there is a table, your perceptual experience as you immediately see a table, and the fact that there is a table. According to Sosa, in a case like this, "visual experience mediates between the fact seen and the perceptual belief through which it is known" (Sosa 2007a, p.46). Now, suppose you have an intuition that a triangle is a three-sided figure. Here we have to distinguish between two things: your belief that a triangle has three sides and the fact that a triangle is a figure composed of three sides. According to him, the perceptual model holds that "intellectual seeming is here seen as a state of awareness with a mediating role analogous to that of visual experience in visual perception" (Sosa 2007a, p.47). But,

[6] Although the intuition-perception analogy was accepted by Bealer in the sense of seeming presence, Yet, Sosa does not agree with this level of analogy simply because it leads us to confuse two different things: seeming and experience. Intellectual seeming is not analogous to experience but to sensory seeming.

in his view, this analogy is misguided. For while it is true that in perception, "sensory experience … mediates between object seen and perceptual belief" (Sosa 2006, p. 208). In intuition, "nothing ... mediate[s] ... between facts known intuitively and beliefs through which they are known" (Sosa 2007b, p. 51). This is not the only functional role experience plays for perceptual belief.

Experience works as a foundational justification for perceptual belief. A foundational justification is "a mental state that provides justification without requiring it in turn" (Sosa 2007a, p.49). Experience, as an irrational basis, then serves a justificatory role for perceptual belief, but it itself does not need justification, for it is a regress-stopper. In Sosa's words, "experiences are supposed to be rationally passive so as to lie beyond rational justification or unjustification. They just happen to us, independently of our rational agency, which is why they are not rationally assessable" (Sosa 2014, p. 41). Here, it is worth noting that although he draws a clear distinction between experience and seeming, he does not deny that there is a perception-intuition analogy. He writes that "the way intuition is supposed to function in epistemology … is by analogy with the way observation is supposed to function in the natural sciences" (Sosa 2007a, p. 106).[7] Sosa discusses what might be the analogous role in the case of intuitive knowledge to the justificatory role that experience plays for perceptual belief. It might be thought that intuitive seeming or attraction to assent does so (Sosa 2014, pp. 41–42). He argues that the perceptual analogue of intellectual seeming is not sensory experience, but it is perceptual seeming.

For that reason, it is worth noting that he does not argue against the idea of seeming. Quite the contrary, seeming is there even in perception. But seeming itself, unlike experience, is evaluable and does not serve as a regress-stopper; rather, it itself needs a foundation as such. In a Muller-Lyer illusion, says Sosa, seeing the lines prompts a prima facie, direct, or "one-thing-considered" seeming that they are incongruent, even if ultima facie, episodic, "all-things-considered" or resultant seeming makes the lines seem to be congruent (Sosa 2007a, pp. 50–51). The initial visual attraction to think the lines are incongruent remains even after a measurement that makes them seem to one congruent. So even in the case of perception, argues Sosa, seeming is operative. The difference is that perceptual seeming is based on experience as a sufficient regress-stopping reason. Concerning intellectual seeming, he holds that although it is not based on experience, it is not baseless. The question now is what foundationally and rationally underlies seeming, what that plausible regress-stopper for intellectual seeming could be?

He argues that a person's sheer understanding foundationally underlies seeming understanding without any "unaided reflection," where reflection includes relevant perception, memory, introspection, inference, and reasoning (Sosa 2003, p. 557). He argues that intuitive seeming is based on nothing other than one's transparent understanding of its propositional content represented by concepts. "I am attracted to think that nothing is self-diverse simply through understanding that content well

[7] It seems that he does not argue against the intuition-perception analogy. Quite the contrary, he argues for the right form of it, according to which intuition should have a similar functional status as perception.

enough" (Sosa 2014, p 42). Understanding is not only ground but also the only foundational justification for seeming. That being so, intellectual seeming, like perception, seems to involve a foundation or regress-stopper of the sort defined.

Now, does understanding qualify as a rational regress-stopper for seeming? Yes, answers Sosa. But in virtue of what, does understanding-based seeming qualify as a rational attraction? Sosa argues that an attraction can also be grounded on a bias, superstition, social pressure, brainwashing, hypnosis, etc. And none of these flawed sources is reason-based. In a case as such, even if the relevant proposition is quite well understood, mere understanding is likely a poor reason (Sosa 2014, p. 44). For that reason, we need to specify a condition that, in combination with understanding, will plausibly explain the rationality of attraction to assent. To that end, he buckles down to the Cartesian, factive model according to which we directly grasp the clear and distinct seeming by appeal to the introspection of an attentive mind. On this view, "a propositional content about a present state of consciousness can attract assent through its sheer truth" (Sosa 2007a, p. 57). Sosa draws two main motives to refute this model. "One problem is shared by this appeal to truth with the earlier appeal to understanding, for truth per se is no less ludicrous as a source of justification than is understanding per se, yet no condition has been specified that in combination with truth would plausibly explain foundational justification" (Sosa 2007a, p. 57). Another problem is that if this direct grasp model is true, then it is defendable through fallacious reasoning, as in the false but intuitively justified attractions and the fallacy of affirming the consequent (Sosa 2007a, p.57–8). The same story goes for the appeal to simple truth as a condition that explains foundational justification. For "no proposition will ever be that much simpler than its negation" (Sosa 2014, p. 46). As a consequence, he concludes that neither distinctness nor simplicity suffices on its own to explain what makes an attraction rational.

Instead, Sosa argues for intellectual competence as a source of rationally structured attraction (Sosa 2013). The competence model is a particular application of virtue epistemology to the epistemology of intuition in order to locate intuition within the scope of an a priori theory of knowledge. This view consists of two main parts. One is that intellectual seeming is intuitive when our attraction to assent a propositional content is drawn by considering it consciously with understanding, as in the example "that 2 + 2 = 4; that no cube is a sphere; that nothing is numerically self-diverse" (Sosa 1998, p. 262). Since our assent can be attracted only by an unaided understanding of the propositional content represented by concepts, this intellectual competence guarantees that our conceptual ability is exercised in our understanding of such propositional contents. Another is that such an attraction is rational because it is explained by competence as such. When one intuits, one is epistemically virtuous because one performs one's epistemic virtue of intuition. In view of that, intuitive seeming is rational if and only if it is given rise by such competence present in one's attraction to assent. And then Sosa's point is that intuition is competent or competence-based if and only if it is derived from such intellectual competence, and it is basis-dependent or incompetent if and only if it is derived from bias, superstition, social pressure, perception, memory, etc.

In sum, we have arrived at a general account that intuition is intellectual seeming. And intellectual seeming is an attraction to assent to a propositional content based only on consciously understanding it. And such an attraction is rational if and only if it is explained by epistemic competence, ability, or virtue. Unlike intellectual seeming, a crucial feature of sensory seeming is that it is an attraction to assent on the basis of both one's understanding as well as a sensory experience. That being so, Sosa thinks that he avoids the challenges he raised against the previous two models, namely, the perceptual model and the Cartesian or introspective model.

However, although understanding plays a crucial role in his account of the nature of intuition, what we are told by Sosa about what it is to understand propositional content is very little. To make the point more precise, let us recall his understanding-view of intuition. For him, intuition is intellectual seeming, which is, in turn, an attraction to assent a propositional content on the basis of the subject's understanding of it alone. Let us consider a type of example that arises from the application of this view. In the 10 coins-case, Gettier's intuition that 'Smith does not know that the one who will get the job has ten coins in his pocket' is intellectual seeming which attracted Gettier to assent to its propositional content as true, merely because he understood it. It is safe to say that beyond a statement of this account, Sosa did not develop an argument that explains how a person's understanding of propositional content alone can serve as an epistemic ground to make that person attracted to assent to that very propositional content.

Further, an opponent of his view may argue that this proposal is an area of some controversy. Such an opponent may say that I may understand many epistemic propositions without being attracted to assent to any of their propositional contents. I may understand many epistemological theories that were given out as propositions without being attracted to assent to any of them. Moreover, I may equally understand two incoherent propositions, an opponent may say, yet I may be attracted to assent to the propositional contents of one of them. Suppose I equally understand both the traditional conception of knowledge according to which 'knowledge is justified true belief' and Gettier's view according to which 'knowledge is not just justified true belief', yet I may only be attracted to assent the propositional content of the proposition 'knowledge is justified true belief.'

Moreover, suppose for the sake of argument that a person's understanding of a propositional content alone makes that person attracted to assent to the very propositional content represented by concepts. Simply put, intuition is based solely on conceptual competence. Accordingly, the word 'understanding' means grasping the concepts that represent the propositional content. Take, for example, Gettier's intuition that 'Smith does not know.' On this reading of the understanding-based view, understanding of the propositional content of this intuition consists in grasping its concepts expressed by "Smith," "does," "not," and "know." I do not think that grasping these concepts is enough to assent to the propositional content of intuition. As Nagel writes, "popular presentations of fundamental scientific discoveries, given out as propositions to which one must subscribe without really understanding them. For example, people are now told at an early age that all matter is really energy. But despite the fact that they know what 'is' means, most of them never form a conception of what makes this claim true" (Nagel 1979, p. 177). Rather, things other than grasping these concepts may be needed, such as their mode of combination. Let us assume that a proponent of the understanding-based view says that by 'understanding,' it meant both grasping concepts and their mode of combination in a proposition. Still, understanding does not qualify as a compelling reason for assenting. For to understand how 3 follows 2, as Wittgenstein argues, is just to learn one technique. It is just a technique of counting that we were trained in.[8]

2.2.3 *Chudnoff's Intellectual Perception Analysis*

Like Bealer, too, Elijah Chudnoff sets out a phenomenologically modelled account of intuition according to which intuition is "a form of intellectual perception" (Chudnoff 2013a, p.1).[9] This account is an improvement, and support of a view that has a long tradition in epistemology from Plato to the present. However, he characterizes it in terms of three similarities and two differences between intuition and perception. Similarities lie in that [1] both intuition and perception are phenomenologically unique experiences, [2] these experiences directly justify beliefs, and [3] They enable us to be aware of their subject matter and accordingly acquire knowledge.[10] Differences lie in that intuition experience can happen involving perception experience, and it may happen alone, and both include their subject matter. However, while the subject matter of perception is a concrete reality, the subject matter of intuition is abstract reality (Chudnoff 2013a, p. 3). By 'abstract reality' he understands "the necessary, normative, infinite, and abstract as in non-spatiotemporal and causally inert. Mathematics, metaphysics, and morality are about it" (Chudnoff 2013a, p. 11). To put it simply, in perception, "the ... relationship [is] between a concept and a particular object". A concept is an ingredient that is linked to an object. In intuition, "it is a relationship between concepts" (Chudnoff 2013a, p. 5). Concepts are both ingredients and objects. He attempts "to motivate the view by situating it within an approach to a priori knowledge," according to which "our knowledge of abstract matters, or at least our knowledge of some abstract matters, is based on purely intuitive reasoning—i.e., reasoning that is both a priori, so not empirical, and intuitive" (Chudnoff 2013a, pp. 13–6). The question now arises: what is it for a person to have an intuition?

---

[8] Further considerations of this proposal are left until Sect. 2.3.

[9] The view that intuition is a phenomenologically unique experience has a lot of believers, such as Pollock 1974; Plantinga 1993; Huemer 2005b; etc.

[10] Intuition is experience: he shares this idea with Bealer, who thinks that intellectual seeming is a state of awareness or a consciously cognitive and reflective episode. Sosa does not agree with this view because it presupposes that intuition has a mediating role analogous to that of visual experience in visual perception. For Sosa, this analogy is misguided: sensory experience mediates between the object seen and perceptual belief, but nothing mediates between facts known intuitively and beliefs through which they are known.

In order to answer this question, he starts challenging doxasticism about intuition, according to which "intuitions are … doxastic attitudes or doxastic dispositions," and defending sui generism, according to which "intuitions—like perceptual experiences—are pre-doxastic experiences that—unlike perceptual experiences—represent abstract matters as being a certain way" (Chudnoff 2011a, p. 626, 2013a, pp. 25–6).[11] He offers two anti-doxastic arguments. One is "that intuition experiences do not imply the presence of doxastic attitudes or dispositions" (Chudnoff 2013a, p. 17). That is, one may intuit without judging or having any inclination to judge. For example, in perception, you might have a perceptual experience that Muller-Lyer lines are not the same in length without judging or having any inclination to judge that they are different in length. Likewise, in intuition, it is possible for you to have an intuition that Naïve Comprehension Axiom is true without judging or having any inclination to judge that it is true. If so, if intuition is not judgment or inclination to judge, "it is a sui generis experience. That is: in having an intuition experience representing that p, it intuitively seems to you that p, and this is a sui generis experience, irreducible to other experiences" (Chudnoff 2013a, p. 44). The second argument is "that doxastic attitudes and dispositions do not imply the presence of intuition experiences" (p. 17). That is, one may judge or have an inclination to judge without intuiting. The major point of this argument is that both intuitions and perception possess presentational phenomenology but judgments or inclinations to judge do not.

In (Chudnoff 2012a), his main aim is to answer the question of what is the nature, scope, and significance of this presentational phenomenology. Regarding its nature, he holds that the experience that has presentational phenomenology both represents the relevant proposition and makes it seem as if the subject is aware of its truthmaker. His idea of the scope of an experience as such can be stated as follows: these very considerations are equally applicable to at least perceptual, intuitive, introspective, imaginative, and recollective experiences. To lend more plausibility to this view, he argues that, in perception, you might judge or have an inclination to judge that the scene is a certain way without—presentational phenomenology—you feel the presence of the scene's objects and their features directly to your mind (Chudnoff 2013a, p. 31).

Similarly, in intuition, there is a difference between judging or having the inclination to judge on the one hand and "entertaining the proposition that diameters determine lines of symmetry, and intuiting that diameters determine lines of symmetry" on the other hand (Chudnoff 2011a, p. 637). The difference lies in that in the latter case, the subject will have a feeling of presence in the mind of the properties of being a diameter, line, and symmetry, but in the former case, the subject won't have such a feeling. Therefore, he concludes that "intuitions have presentational phenomenology and that judgments as well as inclinations to make judgments do

[11] If Bealer accepted the intuition-perception analogy in order to defend the presence of seeming and if Sosa accepted it to clarify the functional role of intuition, it is Chudnoff's main strategy to defend his view of intuition as intellectual perception.

not" (Chudnoff 2013a, p. 52). More importantly, Chudnoff, in the context of presentational phenomenology, explicitly refers to Husserl.

In fact, presentational phenomenology plays a central role in Husserl's epistemology. Chudnoff borrowed this Husserlian basic idea and developed it to be a distinctive aspect of intuition. He writes: "all intuition experiences have presentational phenomenology with respect to at least part of their content." This is what he calls a 'presentationality' thesis. Presentational phenomenology does two things. One is that it makes the intuition seems to its subject that it is the case. Another is that it makes the subject aware of how such intuition is the case. He also introduces what he calls a 'constitutedness' thesis, according to which "all intuition experiences are constituted by other experiences, e.g. conscious thoughts, imaginings, etc." though an intuition experience is distinct from its constituents (Chudnoff 2013a, p. 58). He considers some worries about these theses.

Regarding presentationality, he answers the question of whether all intuitions have presentational phenomenology. His response is yes, but he adds that there are reasons why some intuitions apparently look as if they do not have presentational phenomenology, such as (1) confusing intuition experiences with intuitive beliefs, which sometimes might lack presentational phenomenology that; (2) in some cases might be difficult to catch or; (3) in other cases might be hard to describe and; (4) sometimes "some intuitions lack presentational phenomenology with respect to some [Not all] of their content" (Chudnoff 2013a, p. 60). But presentational phenomenology is omnipresent; that's as plain as the nose on someone's face.

Regarding constitutedness, he considers a Cartesian worry that some intuitions lack imagination; they are of pure thought and, accordingly, lack presentational phenomenology. Giving many mathematical examples, Chudnoff replies that "there is no obvious conceptual confusion in the idea that some experiences both have presentational phenomenology and consist wholly of thoughts" (Chudnoff 2013a, p. 61). All reflections, such as thought, imagination, etc., coincide with intuition experience that sometimes persists after them—where reflections are memory, for example. Even in the case of an unreflective experience, "intuition experience … possesses presentational phenomenology with respect to [minimal parts or] more than minimal parts of its content" (Chudnoff 2013a, p. 72). This is what happens with a subject having logical, mathematical, or philosophical intuition.

We have come to the following conclusions: an intuition, according to Chudnoff, is an a priori sui generis experience or mental state which is constituted by other experiences, e.g., conscious thoughts, imaginings, etc. And such an experience possesses a presentational phenomenology that does two main things: it makes the relevant intuition seems it is the case; it makes the subject aware that such intuition is the case.

As we have seen, Chudnoff's basic strategy to defend this view of the nature of intuition places so much weight upon an assumed analogy between intuition and perception. As I understand Chudnoff, when I see my grey laptop in front of me, it visually seems to me that there is a grey laptop. This is an example of perceptual seeming. When I realize that $1 + 1 = 2$, it intellectually seems to me that $1 + 1 = 2$. This is an example of an intellectual (intuitive) seeming. To be precise, his account is equally as much about perception as intuition. The idea of the intuition-perception analogy is not new. For example, Robert Audi writes, "intuitions are analogous to visual beliefs grounded in seeing, which also provides non-inferential but defeasible justification for those beliefs" (Audi 2008, p. 485). Many other philosophers also have widely discussed the issue at hand.

To turn back to Chudnoff's analogy, it easily raises many worries that are not easy to account for. His argument goes as follows. Two things are analogous if and only if they are phenomenologically so. Intuition and perception are phenomenologically analogous. Therefore, they are analogous. The problem lies in the first premise. Let us suppose that intuition and perception are phenomenologically analogous. Yet, there are no conclusive or compelling reasons or cohesive arguments to reduce their differences to a supposed unique phenomenology. There are so many things that apparently look analogous, but they are not. If plain water and Eastern Mediterranean white alcohol (Arak) phenomenologically seem similar, it does not mean they are analogous; their structure is not alike. Since their structure is not alike, they do not function likewise, and accordingly, they are two different things. I am not arguing that intuition and perception are not phenomenologically analogous.[12] What I am trying to say is that even if they are, their phenomenological analogy may not be enough to warrant their analogy. For someone may object that what makes two mental states analogous is not their phenomenology but the nature of the process that gives rise to them. In this case, we are required to check whether the cognitive processes that give rise to the idea that intuition and perception are analogous are themselves similar. If yes, they do function likewise, and accordingly, they amount to the same result (result = their capability to justify).

Chudnoff cannot easily close his eyes to these points. It is open to him to keep defending his view about the phenomenology of intuitions. But unless he meets this challenge, defending his view on the basis of the aforesaid analogy becomes difficult. This is not to say that saying that intuition comes with a certain phenomenology is incorrect because the assumption of the intuition-perception analogy fails. The claim of intuition's phenomenal character might be correct even if the view of the intuition-perception analogy fails.[13] It is not necessary for the former to be established on the latter. The former might very much be established on something else or even on its own. But in that case, Chudnoff has to find a way other than his strategy of intuition-perception analogy to defend the phenomenology of intuition.

I think that this is a good entry to go further. One might ask, what if intuition cannot be characterized in terms of a supposedly special phenomenology? What if the authors of the aforesaid ten thought experiments were asked about the special phenomenal character their intuitions have, and they replied that they have none? In that case, why should we trust the claim that intuition can be demarcated in terms of phenomenology at all? These questions might seem rather extreme. Nevertheless, they have been asked by some philosophers. Herman Cappelen, for example, in

[12] Though I will do so in Sect. 2.3.

[13] Though I will be arguing to deny a phenomenal character as such later in this chapter.

Philosophy without Intuitions (2012), thinks that it is just a common mistake of philosophers to suppose that intuition has characteristics such as having a special phenomenology, etc. Here, I do not intend to support Cappelen's argument against Chudnoff's. All I want to do is to know on the basis of what reason Chudnoff claims that intuition has the putatively characteristic phenomenology. If I understood his argument correctly, I think the only answer he has is that intuition is like perception. Perception has the character of phenomenology. Therefore, intuition also has the character of phenomenology. But what if such an analogy is just like the Water-Arak one? In a case like such, Chudnoff is required to provide us with a more plausible response.

What is more, Chudnoff asserts that the phenomenology of intuition is structured by other conscious experiences such as thoughts, imaginings, etc. But he completely rejects the reducibility of the phenomenology of intuition experience to these ingredients. To support this view of irreducibility, he gives an example that "a statue is distinct from but constituted by the clay out of which it is made" (Chudnoff 2013a, p18). Here, there is something missed. The clay and the statue are distinct not only in the sense of form. But they are also distinct in the sense of space. That is, the amount of clay used in making the statue did not only take a new form. But it also occupied a new space. An opponent of Chudnoffism might tend to worry about whether or not the phenomenology of intuition experience occupies a new and distinct space of the land of consciousness from that of its ingredients. But he gives no strong rebuttal or confirmation of this query for the independent form and space of the phenomenology of intuition existence; indeed, he does not discuss this issue at all, at least not to my knowledge in his works. I don't mean that this is a fault of Chudnoff's. Chudnoff gives no clear indication as to what he takes to be the form and space of the phenomenology of intuition existence or how this phenomenology is to be determined. But that may be because he is writing on the phenomenology of epistemology and not the metaphysics of epistemology.

Moreover, a different challenge arises against Chudnoff's account of the phenomenology of intuition experience. For it is hard to see how it is possible for different intuitions to share the putatively characteristic phenomenology, which is made by different conscious experiences. "Why should different intuitions have something phenomenological in common if they are constituted by very different collections of mental states?" (Koksvik 2017, p. 5). If the phenomenology of X intuition was made by imagination and introspection, and the phenomenology of Z intuition was made by inference and perception; how is it possible for X and Z to share the same phenomenal character of being spontaneous, for example?[14]

[14] I am leaving further considerations of this proposal until Sect. 2.3.

2.3 Ascriptions of Intuition

Each of the preceding accounts expands the analysis of the nature of intuition. Intuitions were considered mainly to be: knowable a priori, propositional attitude (i.e., a mental state held by a subject toward a proposition), and phenomenal (i.e., what it is like to be intuiting). Maybe the previous accounts attributed more features to intuition. However, in what follows, I will be focusing mainly on these three conditions. Let me address each in turn.

2.3.1 Intuition and the A Priori

The topic of a priori played an important role in the history of epistemology. The historical source for contemporary debates on the a priori has its origin in Kant's *Critique of Pure Reason*. Kant articulates the term by contrasting it with the a posteriori. He draws the a priori-a posteriori distinction by linking it to the cognitive source of the justification. According to him, if knowledge is wholly independent of all kinds of experience, then it is called a priori.[15] In contrast, if knowledge is derived from any type of experience, then it is called empirical or a posteriori. He writes: "we will understand by a priori cognitions not those that occur independently of this or that experience, but rather those occur absolutely independently of all experience. Opposed to them are empirical cognitions, or those that are possible only a posteriori, i.e., through experience" (Kant 1998, p. 137). Of course, this approach implies a negative program: knowledge is a priori only if its justification is free from any experiential aspects; otherwise, it is a posteriori. As a result, neither there is a third option apart from the a priori and a posteriori options, nor is there any borderline case between them.

However, the Kantian concept of intuition (Anschauung) benefited from this distinction.[16] In light of the above distinction, Kant demarcates a semantic distinction to distinguish between two kinds of truth: analytic truths and synthetic truths. Analytic truths are reduced to the meaning of the words and concepts involved. This

[15] It is important to emphasize that this is what the Kantian a priori means. However, there are various other accounts of the a priori. For example, Foucault argued for a historically constructed sort of a prior. According to The Cambridge Foucault Lexicon, "the historical a priori serves as a mechanism by which certain specific, bounded, and concrete practices (like the gaze, which organizes aesthetics) are able to saturate and reorganize domains seemingly far removed from them (domains like medicine). And, in the process, these "historical" discourses sculpt new "a priori" objects for their disciplines: they construct (in the guise of "discovering") new objects and new protocols that enable different engagements and new methods of investigation" (Nealon 2014, p. 201). Also, C. I. Lewis developed a pragmatic/relativized account of the a prior, according to which "a priori knowledge arises from the analysis of conceptual principles: practices that are nested in the conceptual scheme we are, by convention, committed to" (Järvilehto 2011, p. 107). Thanks to an anonymous referee for raising this concern.

[16] Thanks to an anonymous referee for raising this concern.

can be simply seen in the classic case of "all bachelors are unmarried men." The subject (bachelor) contains the predicate (unmarried man). Synthetic truths are based on both the contents as well as empirical confirmation. For example, the geometrical truth "a straight line is the shortest distance between two points" is synthetic because "straight" is not synonymous with quantum. Thus, this proposition, if true, cannot only be known independently of all experience. It's truth, and all kindred truths can be known better through investigation.

What is at stake for Kant are the foundations for any claim of knowledge of the external world. To that end, Kant developed his transcendental arguments to delineate the necessary conditions for the possibility of experience. Briefly considered, sensory experiences must be spatiotemporally framed. In a transcendental mode of analysis, we must conceive of intuition (Anschauung) as featuring a disposition to apprehend the external world. As is a disposition, intuition is a priori, albeit it can be given in the empirical. The pure forms of intuition, namely, space and time, structure the experience. That is to say that they are themselves a priori, but they are given in the a posterior. If so, it follows that although intuitions of space and time are nonempirical, they can ideally establish synthetic truths.

In the last few decades, however, epistemology has witnessed interesting developments that bear directly on the a priori–a posteriori distinction.[17] Intuition is an example of developments as such. It was often said that it is intuition that justifies Jackson's thought experiments of Fred and that of Mary, Searle's Chinese Room case, Chalmers' Zombie case, Cohen's two Lottery cases, Lehrer's Truetemp case, and so on. In view of that, these developments led to investigations of the nature of this justification or intuition. This brings up the question: is intuition a priori or a posteriori? At this point, some epistemologists try to situate their views of intuition within the scope of a positive approach of a priori justified knowledge. Here is Bealer, for example, who says very clearly that when he speaks of intuition, he means a priori intuition (Bealer 1999a, p. 30). In like manner, Chudnoff attempts to show the importance of his view of "intuition as intellectual perception … motivating interest in it by relating it to the broader topic of the a priori" (Chudnoff 2013a, p. 1). In a similar fashion, Sosa holds that intuition is "involved in abstract, a priori, armchair thought" (Sosa 2007c, p. 101). But how do these theorists argue for the a priori intuition?

Bealer, Chudnoff, and Sosa wholeheartedly follow the Kantian conception of a priori while speaking about intuition. Bealer argues that philosophy answers its questions by appealing to intuitions, where sense experience is thought to be irrelevant. He argues that philosophy is autonomous from empirical sciences such as cognitive psychology, psychology, neuroscience, and so on. In case of conflict between philosophical and scientific answers to a certain philosophical question, he asserts that intuition outweighs empirical data, and accordingly, philosophy has authority over empirical sciences. He writes that:

---

[17] Casullo and Thurow restrict those developments to intuitions, the recent movement of experimental philosophy, the concept of possession, the new disagreement problem, contextualism, and naturalism. For more details, see (Casullo and Thurow 2013).

> all philosophers seek answers to such questions as the nature of substance, mind, intelligence, consciousness, sensation, perception, knowledge, wisdom, truth, identity, infinity, divinity, time, explanation, causation, freedom, purpose, goodness, duty, the virtues, love, life, happiness, and so forth (Bealer 1998a, p. 203).

Then the central terms of philosophy are relevant to metaphysics, philosophy of mind, epistemology, ethics, logic, and philosophy of science. This seems to me to be close to Chudnoff's view that dichotomizes a priori vs. a posteriori reasoning as follows. In empirical reasoning, sensory perception injects a posteriori reasoning with some content. In intuition, some conscious episode does provide a priori reasoning with some content. He writes that "there should be some conscious episode that, just as sensory perception does for empirical reasoning, injects a priori reasoning with some content. Intuitions thought of as intellectual perceptions fit the bill" (Chudnoff 2013a, p. 16).

In the same vein, Sosa analogizes the role of intuition in a priori knowledge to the role of perception in empirical knowledge as follows. He argues that seeming is operative both in perception and intuition as well. However, he distinguishes between intellectual and perceptual seemings. Both are based on conscious understanding. Nonetheless, perceptual seeming is based on conscious understanding as well as a sensory experience. Conscious understanding alone does not yield attraction. Only when conscious understanding combines with experience can attraction be exerted. While, in intuition or intellectual seeming, an unaided conscious understanding of a propositional content is enough to yield an attraction to assent (Sosa 2014, p. 47). Thus, intuition is understood to be a source of a priori knowledge.

In light of the above discussion, according to Bealer's, Chudnoff's, and Sosa's version of the Kantian conception of a priori, intuition is best understood as a source of a priori knowledge. This is what is commonly rejected by experimental philosophy—a very recent development—that appears to clash with recent epistemological literature, offering an entirely different view.

2.3.1.1 The Experimental Approach[18]

One of the most salient recent developments in epistemology is the experimental philosophy movement. Waves of work have been written by proponents of this movement, explaining what intuition is like. It was often said that the results of experimental philosophy constitute a serious challenge to the view that intuition is a priori. Whether these results affect the a priori status of intuition is an interesting

[18] It is worth noting that the recent development of experimental philosophy is an approach that makes use of the methods of experimental psychology and cognitive science to come up with something new regarding both the traditional methods and claims about issues in philosophy. However, it can be divided into two programs: negative, which questions the traditional methods of analytic philosophy, and positive, which props up the conventional claims about traditional philosophical matters.

question about the a priori. In the present section, I shall be presenting the argument from experimental philosophy.

When all is said and done, experimental philosophy has advocated the thesis that the apriorists' approach is questionable. Experimental philosophers argue that an epistemic intuition about thought experiment is a source of a priori (non-experiential) knowledge if and only if it is intuited by a subject without looking into the world, where the word "world" includes effective variables that are irrelevant to the matter the thought experiment seeks to address. To investigate this premise, they conduct pilot studies surveying the intuitions of both non-philosophers and philosophers about thought experiments that are similar to those that philosophers have employed in their texts. Making use of the methods of experimental psychology and cognitive science, proponents of this movement claim that their results show that intuitions are affected by factors that are irrelevant to the issues the thought experiments aim to address. For instance, Gettier's intuition that Smith does not know could be influenced by factors irrelevant to the concept of knowledge. The followings are examples of these studies.

Among the examples of such variables, the results of a recent investigation of epistemic intuitions by experimental philosophers talk about ethnicity (Nichols et al. 2003; Feltz 2008), socio-economic status (Weinberg et al. 2001), cultural dissimilarities (Weinberg et al. 2001; Machery et al. 2004; Nichols 2004; Spicer 2010); educational background (Kuntz and Kuntz 2011; Weinberg et al. 2001), native language (Vaesen et al. 2013), gender (Buckwalter and Stich 2013; Adleberg et al. 2015; Seyedsayamdost 2015), the order in which the thought experiment is presented (Swain et al. 2008; Weinberg et al. 2001), individual differences (Nichols and Ulatowski 2007), age (Colaco et al. 2014), moral judgments (Knobe 2010), and so on and so forth. The proponents of experimental philosophy have concluded that this data raises serious questions about the a priori status of epistemic intuition.

In other words, experimental philosophers argue that the results of their studies constitute experimental evidence that puts the apriority of intuition in deep trouble. Being affected by these factors, intuitions are now considered cases of our cognition of the empirical world. However, some philosophers assimilate experimentalists' findings and argue for the a priori status of intuitions. Their arguments for doing so will be our concern in what follows.

2.3.1.2 Peacemaking Attempts

In recent years, groups of philosophers have taken on the roles of reconciling adversaries, those who believe that intuition is not a priori bolstered by the results of experimental philosophers and those who believe that intuitions have some a priori purchase. They think that apriorism has to and can assimilate with the empirical findings. That is to say that one has not to attempt to give up apriorism initially because empirical findings can fit in with apriorism. In what follows, I will be laying out the best-known four approaches that try to rescue the apriority of intuition from the impacts of experimental philosophy attacks. I shall also point out the reason why I think these arguments are not satisfactory. Here are the arguments.

2.3.1.2.1 Conciliatory Argument

Philosophers have made great efforts to conciliate in the dispute. For example, (Goldman 2013) nicely distinguishes between two kinds of questions about intuitions. One is the first-order question of whether intuition qualifies as evidence at all. Another is the second-order question, that is, is there good a priori or empirical evidence that says intuition qualifies as evidence? In light of these, he reads experimentalists' findings as second-order evidence about intuitions' first-order evidentiality. In other words, the results of experimental philosophy provide second-order empirical evidence against a priori first-order evidence of intuitions. Therefore, intuition is a priori evidence, but whether there is evidence indicating its evidentiality is an empirical task.

In the same spirit, (Ichikawa 2013) denies the contention that the results of experimental philosophy, even if they are correct, undermine the a priori status of intuitions. He maintains that experience can play an enabling role in one's having an a priori intuition, but it cannot warrant such intuition. Experience facilitates one's having an a priori intuition, but it cannot be part of its warrant. For example, experience enables Jackson to intuit that our having all physical information about the subject won't help us to know everything about her color experience. But experience does not warrant this intuition. For more explication, suppose that Jackson thinks he has no good reason to have the following a priori intuition: all physical information about the subject won't help us to know everything about her color experience. Call this intuition Z. A thought as such defeats the apriority of his intuition Z. Yet, a new experience may present evidence that he does not lack a good reason to a priori intuit Z. This experiential evidence defeats the previous defeater and saves the a priori character of his intuition Z. Nevertheless, whether this intuition is warranted or not is a different question. Accordingly, for Ichikawa, the apparent inconsistency between empiricism, according to which one has to proceed from empirical data in order to get knowledge, and apriorism; according to which one has to begin one's course of obtaining knowledge from the armchair, is just an illusion. And, experimental philosophy, if true, has just a limited effect on armchair intuitions.

This conciliatory approach is also defended by (Thurow 2013). Thurow introduces a theory that maintains both the a priori character of intuitions and the viability of experimental philosophy. On this view, the a priori character of intuitions consists in their being comprised of concepts. And the viability of experimental findings is preserved since the content of concepts is determined by one's community. Like Ichikawa and Thurow, (Weinberg 2013) argues that experimental results may modify the conception of the a priori intuition. Nevertheless, an intuition as such can correctly be deemed a priori in the sense of Ichikawa's view of defeating the empirical defeaters. In other words, these intuitions are a priori, but not completely so. In this sense, experimental results can cause only minor modifications to the notion of a priori intuition.

In accordance with Weinberg, (Henderson and Horgan 2013) argue for "low-grade" kind of apriority. In short, an a priori judgment is conceptually grounded justification, i.e., it is an analytic judgment because its quality of being a priori is based on the meanings of its constituent concepts. Yet, it has an inescapable empirical feature since its justification is based on the very meaning that is empirically informed by abduction. For example, although the claim that "Water = H20" can be conducted from the armchair by virtue of the meanings of its constituent concepts, its justification abductively comes by way of empirical intuitions, facts, scientific knowledge, and other data. "Water = H20" is analytically generated by an armchair reflection based on the meanings of its concepts. Nevertheless, it involves an ineliminable empirical dimension because its justification is constituted by the content of its concepts which is acquired by empirical abduction. In view of that, the justification of a claim as such may be a priori, but not purely so.

Analyticity is defended by (Hill 2013) against the revisability thesis, according to which any statement is revisable by empirical evidence, a position maintained by Quine. Contrary to that, Hill develops and defends a threefold argument that argues in favor of the analytic statements' immunity to revision. On this view, certain beliefs possess epistemic immunity, teleological immunity, and a combination of the two. The laws of mathematics possess epistemic immunity since we accept them by virtue of the fundamental dispositions their concepts aesthetically serve. The laws of classical logic possess teleological immunity since we accept them by virtue of the fundamental cognitive concerns their concepts serve independently of our interests. Such cognitive concerns cannot be reduced to the idea of mere efficient prediction, as in the case of religious, disjunctive, conjunctive, truth, and modal beliefs. The combination of the previous two immunities provides the subject with considerable knowledge of the correct use of the terms. Therefore, at least in many cases, we cannot revise the analytic statement by empirical evidence. Nor can we derive it from such evidence.

To succeed in conciliating both aprioristic and experimental views, one does not have to include any obvious example of the a posteriori knowledge within the scope of the a priori knowledge. For that reason, (Malmgren 2006, 2013) opposes the plausibility of Tyler Burge's argument that expands the scope of the a priori to include testimonial knowledge that traditionally has been considered an obvious example of the a posteriori knowledge. On Burge's view, a subject has a priori warrant both to her understanding and rationality of her source. These two warranties guarantee the source of testimony that someone knows something a priori. If so, then one can know the same thing a priori depending on the source of testimony. Contrary to that, Malmgren argues that Burge's view offers no good reason for why an agent has to believe any testimonial warrant a priori. And, in case it happens that an agent believed what is said, a testimonial belief as such is inevitably a posteriori.

2.3.1.2.2 Anti- skepticism Argument

Traditionally, philosophers used to turn to the a priori in order to avoid skepticism, the view that knowledge is impossible. In light of that, (Pust 2013) takes up this traditional epistemological problem of skepticism seeking to show how intuition as a priori justification is an unavoidable base to get rid of skepticism. To that end, he defends "The Cartesian Perspective" that introspection and a priori intuition are sufficient to make our beliefs justified. In the same vein, he rebuts the Reidian objection that The Cartesian Perspective arbitrarily reduces justification to introspection and a priori intuition. He turns this objection on its head and argues that unlike other sources of justification, such as perception and memory, at least most deliverances of introspection and a priori intuition are justified. If so, then it is the Reidian account that is arbitrary.

In a quite different way, (Wedgwood 2013) argues that skepticism about a priori justification is misplaced. He explains the skeptical argument as follows. Our beliefs about the world are justified if and only if we are justified in believing that we are not in a skeptical scenario. Only experience can provide us with a justified belief that we are not in a skeptical scenario. Unfortunately, the experience cannot justify such a belief. Wedgwood argues that the only way to respond to the skeptical argument is to accept the fact that we have a priori justification for believing the propositions that the world is knowable by our experiences. This argument relies on the assumption that there is a belief-forming process that considers taking experience at face value to be rational. But this rationality of the belief-forming process is taken for granted. Accordingly, he ends up facing a more fundamental question: how to explain the rationality of a belief-forming process as such.

2.3.1.2.3 Argument from Anti- A Priori–A Posteriori Distinction

Some contemporary philosophers argue that the a priori–a posteriori distinction is defendable. For example, (Casullo 2013) evaluates the challenges to the a priori–a posteriori distinction presented by Philip Kitcher, John Hawthorne, C. S. Jenkins, and Timothy Williamson. Casullo argues that these philosophers failed to fulfill a goal as such simply because they could not articulate a correct concept of the aforementioned a priori–a posteriori distinction. According to Casullo, there are three factors they did not take into account. Consequently, these three factors amount to three errors in articulating the concept. First, they are not clear whether their analysis of the a priori knowledge or justification is theory-neutral, as Jenkins believes, and Hawthorne and Williamson argue, or theory-dependent, as Kitcher thinks. This entails an incoherent articulation of a traditional concept of a priori knowledge within a nontraditional theory of knowledge. Second, they are not clear whether they target the reductive approach to a priori that reduces it to the a priori justification, as Kitcher and Williamson hold, or the nonreductive approach to a priori that expands it to mean the a priori knowledge, as Hawthorne prefers. This makes it hard to know whether the source of the aforesaid incoherence is the a priori itself or the background theory of knowledge. Third, they do not specify a standard that sets apart a priori knowledge or justification from a posteriori one. All they mention is what is common to both. That is why they were not able to distinguish between what is constitutive for a priori knowledge or justification and what is constitutive for the background theory of knowledge or justification. As a result, Casullo concludes that these philosophers have not provided any persuasive reason to deny the a priori–a posteriori distinction.

Nevertheless, the a priori–a posteriori distinction cannot be drawn clearly. (Williamson 2013a) argues that the distinction between a priori and a posteriori way of knowing is superficial simply because there are many borderline cases that are knowable both a priori and a posteriori. An example of these borderline cases could be Williamson's own example of Mary and John, who have come to the same disjunctive truth in two different ways. Mary deduced it a priori, while John deduced it a posteriori. Consequently, cases like these show that a priori and a posteriori are not substantially different, and accordingly, the aforesaid distinction is theoretically of little interest. He does not deny a priori knowledge. Nor is he saying that knowledge is always a posteriori. All he is trying to say is that both are there. But the distinction between them is shallow to the extent that it cannot be drawn.

2.3.1.2.4 Argument from the A Priori-Naturalism Compatibility

Some argue that the idea of a priori intuition is incompatible with naturalism, the view that the world is what science says it is. On the contrary, (Jenkins 2007, 2008, 2013) resists W. V. O. Quine's, Penelope Maddy's, and David Papineau's thesis that the a priori knowledge or justification clashes with naturalism. Quine argues that if empirical science does not confirm an a priori faculty that gives rise to small analytic (concept-sized) units of meaning, then the acquisition of a priori knowledge is not possible. To this, Jenkins replies, we get concepts in two ways. Either we acquire concepts in response to sensory input, or sensory input filters out pre-existing concepts. If so, then the meaning is not too far separated from what experience can provide. Furthermore, Maddy holds that adopting naturalism as an empirical scientific enterprise leaves no space that can be occupied by a priori knowledge. In her response, Jenkins states that it is true that naturalism leaves no space for a priori separately from the empirical scientific enterprise. Yet, this offers us no reason to say that both enterprises cannot coexist. Finally, Papineau holds that if philosophical and scientific inquiries are the like, then a posteriori knowledge, not a priori, plays a crucial role in philosophy. For example, we do not trust intuitions that were often strongly associated with aprioricity until they get a posteriori justification. Nevertheless, Jenkins meets this challenge by arguing that Papineau seems to be interested in the possibility of error more than he is when it comes to the aprioricity. However, intuition is not the only source of a priori in philosophy. As a result, Jenkins concludes that the thesis of a priorism-naturalism clash fails.

I find these arguments for the a priori-a posteriori coexistence interesting. But they still look at the issue through the traditional terms, while they have enough reason not to do so. If intuition may be knowable a priori, a posteriori, both a priori and a posteriori, or neither a priori nor a posteriori, that means it is neither a priori nor a posteriori. It transcends the merely traditional distinction about the ways of knowing. I take that this is what intuition is. So, I am not exactly arguing for a conciliatory position, yet I accept it as a step toward a more ambitious account of the nature of intuition. In what follows, I will start by presenting two anti-apriorism arguments.

2.3.1.3 Two Possible Responses

While it may seem obvious, for some, that intuition is a priori; there do seem to be a few possible arguments that can be put forth to argue otherwise. In this section, I introduce two of these possible arguments against apriorism about intuition. One is against Chudnoff's exemplifying of the a priori. And the other is against Sosa's, Chudnoff's, and Bealer's view that explains a priori in terms of concepts. The former asserts the conciliatory position and concludes that Chudnoff is wrong. The latter introduces empirical evidence that concepts involve empirical elements. Based on these two arguments, I show that apriorism about intuition seems to be untenable.

2.3.1.3.1 Against the Number Sequence Argument

Chudnoff has presented his answer to the question of why intuition is a priori. His response consists in that intuition is part of our abstract knowledge that is purified or free from any sensory or experiential elements. The number sequence is an example of this view. He writes that "I know that every natural number has a finite number of predecessors just by thinking about the structure of the number sequence" (Chudnoff 2013a, p. 15). I agree that examples as such are traditionally considered serious instances of a priori. On the other hand, these kinds of examples can be read in different ways.

Suppose I am not an apriorist. In that case, I wouldn't have any commitments to apriorism. And I might say that our knowledge of examples such as the number sequence is caused by one or all of the following. Firstly, I may argue that they might be caused by memory. They may be stored in the mind that recalls them when it needs them. Suppose I learned the number sequence from my childhood teacher, who taught it to me by using apples. Suppose also I stored in my mind and recalled it once I needed it. Secondly, I may say that they were conveyed to me by a form of testimony. They may be a kind of written or spoken statement given by somebody else. Suppose I grant whatever is said by my father, suppose I grant the number sequence just because it is a written or spoken statement given by him. Thirdly, I may think that they might be a kind of induction. They may be a kind of general law or principle inferred from particular instances. Suppose I induce the law of the number sequence from many particular instances that happened in my or someone else's experience. And lastly, I might say that it is a sort of counting technique we learned through practice, instructions, etc. As Wittgenstein once said, our knowledge of examples as such may be just a result of some technique, we have learned earlier. He holds that the reason why 13 follows 12 is the fact that they are already given this way in the custom or technique of counting we have learned earlier.

If one or all of the aforesaid probabilities is true, then experience in one way or the other is involved in our knowledge of the number sequence. Thus, our abstract knowledge is not totally purified or completely free from all sensory or experiential elements. It is based, at least in part, on empirical reasoning. It is open to Chudnoff or his proponents to argue that if something is a priori in origin and conveyed by memory, testimony, or induction, it will still remain a priori. In this case, it is difficult to accept that such examples were a priori in origin without giving any plausible argument. In the absence of such an argument, one can claim that the so-called abstract knowledge is not too far from what experience provides.

2.3.1.3.2 Against the Argument from Concepts

A common version of the Kantian conception of a priori among Bealer, Sosa, and Chudnoff is that intuition is best understood as a source of a priori knowledge. They declare their approval or support of this view by introducing what is called the argument from concepts.

The idea is that concepts involve abstract (non-experiential) thought. According to Bealer, we may possess concepts in two ways. None of them is a posteriori. The first is that one can possess concepts in a minimal, weak, indeterminate, or incomplete sense, as in Burge's arthritis example, where the subject of Burge's arthritis thought experiment misunderstood the concept of arthritis. The second is that one can possess concepts in a complete, determinate, full, or strong sense, as in the multigon and chromic examples, and most of the central concepts of philosophy where the term expresses some definite concept. Ernest Sosa develops the idea this way:

> S rationally intuits that p if and only if S's intuitive attraction to assent to <p> is explained by a competence (an epistemic ability or virtue) on the part of S to discriminate, among the contents that he understands well enough, the true from the false, in some subfield of the modally strong (the necessarily true or necessarily false), with no reliance on introspection, perception, memory, testimony, or inference (no further reliance, anyhow, than any required for so much as understanding the given proposition) (Sosa 2007a, p. 61).

Not too far from that, Chudnoff looks at intuition as a relationship between concepts that are ingredients and objects. All in all, based on the argument from concept, they set up a framework that intuition is a priori of the sort defined.[19]

Simply put, the concepts-based view of intuitive justification develops the idea that if your intuition that p justifies you in believing that p, it does so because your

[19] The question that arises here is whether the content is somehow already given in our concepts. However, I am not discussing this point here.

intuition is based on your fully grasping the concepts the proposition that p consists of. Take, for example, Gettier's intuition that 'Smith does not know.' In this reading of the concepts-based view, understanding the propositional content of this intuition consists in grasping its concepts expressed by "Smith", "does", "not", and "know." And since concepts are derived from non-perceptual construals, intuition is a priori.

This attunes to (Fodor 1975)'s view of the concept, according to which a concept is considered abstract, amodal, non-experiential, non-perceptual, and mental representations. Fodor postulates that concept possession includes computation. And this, in turn, requires the organism to confirm a relationship between the observed and inferred reward contingencies. This computational operation is not possible without a system that represents the relevant information and carries out the computations. So, the organism has at its disposal a representational system of the requisite power. This can be clearer in the following experiment. The subject in the thought experiment is required to categorize cards that contain either red circles or black squares. If she places the cards that contain red circles in the pile of red circle cards and the cards that contain black squares in the pile of black square cards, she will only then be rewarded. The subject did just that: she properly sorted the cards and thereby got rewarded. For Fodor, this is a kind of inductive operation which goes beyond the experiential data. And since she learned the concepts of a red circle or a black square without a natural language, the system of representation cannot be determined by a natural language; it is genetically determined.

If I correctly understood Fodor's position, he attempts to say that concepts are part of our biological endowment. A knack as such helps us to acquire any concept from the environment. Since they are genetically available to us without looking into the world, concepts do not include empirical construals, and thereby they are a priori. But are they so? In what follows, I would like to produce some empirical evidence that shows why a priori-based view of the concept is false.

The a priori-based view of concepts does not seem to be satisfactory. My main dissatisfaction with this account is its lack of attunement to the current developments in the sciences of the mind. For example, it does not attune to the recent discoveries of cognitive scientists who have identified certain mechanisms that underlie conceptual processing. To proceed further, cognitive scientists persist in debating the issue of concepts. It is true that according to amodal theories such as Fodor's one, a concept is considered an abstract, amodal, non-experiential, non-perceptual, mental and representation (Tulving 1972; Smith and Medin 1981; Fodor 1975; Pylyshyn 1984). But, contrary to that, recently, a modal approach accumulated promising empirical evidence, suggesting that concepts emerged from the modalities of the mind (Lakoff and Johnson 1980; Mandler 1992; Newton 1996; Glenberg 1997; Clark 1997; Barsalou 1999). In particular, the studies of this growing wing of cognitive scientists show that conceptual processing resides in modality-specific systems for perception (Barsalou 1999, 2003, 2009). As a result of this, concepts are not devoid of perceptual materials. Rather, they are, indeed, derived from perceptual construals.

Let us go through some modal findings. (Barsalou et al. 2003) collected many strong empirical pieces of evidence from experimental studies to show the role of modalities in concepts. Studies on, for example, vision, auditory, movement, imagery, gustatory, etc., demonstrate that perceptual variables (e.g., occlusion, size, shape, orientation, etc.) affect conceptual processing. Here is a brief sample Barsalou and his colleagues use as empirical shreds of evidence to support their modal thesis that conceptual processing is affected by perceptual variables:

> In feature listing, a feature is produced less often when occluded than when unoccluded (e.g. participants produce roots less often for lawn than for rolled-up lawn). Participants simulate a referent of the target concept, with features visible in the initial simulation easier to report than occluded features (p. 86).

This example, along with many other examples, endorses the notion that conceptual representations are involved in modality-specific systems for perception.

These multiple instances of evidence resonate with other evidence from cognitive neuroscience. Studies show that the brain has various areas for conceptual processing. Researchers localized those areas in modality-specific systems for perception. Studies by (Martin et al. 2000; Martin 2001; Martin and Chao 2001) confirm that different conceptual processes activate different modality-specific areas. For example, the conceptual processing of color sensations is confined to an area near V4. And the areas that process forms and motion are near brain centers. Analogously, the motor system is the area that processes action. At the visual level, visual areas process visual sensations, and so on and so forth. In short, different types of conceptual processing activate different modality-specific zones.

Most importantly, though, is how this activation is happening. According to (Barsalou et al. 2003)'s neuroimaging techniques, conjunctive neurons in convergence zones constitute a somewhat amodal mechanism for capturing and re-enacting modality-specific states … that conjunctive neurons generally seem to have modality-specific tunings. Furthermore, their primary role might be to reactivate modality-specific states that then support conceptual processing" (p. 87). All this empirical evidence indicates that concepts involve perceptual, experiential cognition. In light of the traditional distinction, such cognition is just not a priori: it is a posteriori.

2.3.1.4 Intuition and Multi-dimensionality

In the previous section, I have shown how concepts can be derived from perceptual construals. A direct result of doing so is that if the concept-based view of intuition is right, if intuition consists of these concepts, then intuition is a posteriori. Nevertheless, as used by philosophers, intuition is used to show whether some cases are instances of the concept in question. For example, Jackson's intuitions are to show that the cases of Mary and Fred are not instances of physicalism. Also, BonJour's intuition in Norman the clairvoyant is anti-externalist in nature. Similarly, Gettier's intuition that Smith does not know is a case in which the traditional concept of knowledge as justified true belief is not applicable. In these cases, intuition is a priori. Jackson, BonJour, Gettier, etc., utilize no empirical steps to have such intuitions. If so, intuition involves both a priori (rational) and a posteriori (empirical) aspects. That is, intuition has a posteriori components when it is explained as being constituted by concepts, memory, induction, testimony, and technical mechanism. It has a priori constituents when it is just a procedure to show whether some cases are instances of a certain concept without looking into the world.

That being said, what should we call intuition that includes both a priori and a posteriori elements? It may be said that we can call it empirico-rational. The problem with this suggestion is that it treats every intuition as having both empirical and rational grounds equally. But this is a misleading perspective because there are some borderline cases, as in introspection, which is neither empirical nor rational.

Like Descartes, one may acquire some propositions by means of a mere rational reflection and totally independently of experience, such as I exist or I am a thinking thing. As Kant presents it, an introspective acquisition as such comes out a priori without further ado, and thus it can be classified along with logical and mathematical knowledge. Nevertheless, in fact, the notion of experience is not well defined for grounding the distinction. In the case of knowledge by introspection, one is justified in believing a belief by appeal to, presumably, a kind of inner seeing or reflection (i.e., inner experience)—as that I am in pain at a certain time or am thinking about Paris. Of course, if we believe, like Kant, that a priori knowledge is knowledge that one has independently of experience, then introspectively acquiring these examples is supposed to be viewed as a posteriori. For this kind of inner seeing or reflection is conceivable as experiential since it involves some sort of phenomenal character. In this sense, knowledge by introspection is viewed as a posteriori. As commonly conceived, introspection then cannot be arranged in the pure a priori or a posteriori category.

If an intuition emerges from it, it will be introspective intuition and, accordingly, neither empirical nor rational. If that is true, I am raising doubts about the contention that intuition is either a priori or a posteriori. As we have seen, intuition may have both labels, and it may have none. As a result, intuition cannot be explained by the traditional a priori-a posteriori dichotomy which accordingly has to be transcended. Intuition is better to be taken as involving several dimensions or aspects. An intuition might have the aspect of being a priori, a posteriori, both a priori and a posteriori, or neither a priori nor a posteriori. If so, we have come to the conclusion that intuition is multi-dimensional.

But suppose that I am right: intuition is multi-dimensional. Why does this matter? For at least two reasons. Most obviously, realizing that intuition is multi-dimensional teaches us something about its nature and provides a constraint on theory. Someone who wishes to account for the nature of intuition must also consider this fact and must ensure that her theory can capture these dimensions. Secondly, the reasoning which shows that intuition is multi-dimensional can also teach us something about the structure of justification. If epistemic intuition justifies epistemic claims, then those who care about the epistemology of intuition as evidence that justifies epistemic claims should, therefore, also care about its nature. An intuition, one might have thought, cannot provide multi-dimensional justification.

Not so, if what I have argued is correct. In the same vein, there is, however, the question of whether intuition is better characterized as propositional.

2.3.2 Non-propositional Intuition

Propositionalism has become almost an orthodox doctrine in the philosophy of mind, language, the epistemology of justification, and so on. On this view, the proposition is the primary part of the meaning of a sentence referring to one or all of the following. Some philosophers consider it to be the bearer of truth, falsity, and nothing else. For instance, Gottlob Frege believed in the idea that being true means being in relation to an abstract entity, i.e., 'Thought.' Thought is the sense of a declarative sentence under the truth and falsity conditions. Thought is imperceptible, can be expressed but not asserted, and can be under and over-determined by sentential content.

Given that, it is not a constituent of the external or sensible world of objects. Nor does it belong to the inner realm of ideas such as sensations, desires, intentions, and so forth. Since it is independent of both aforementioned realms, the only choice left is the graspable but non-spatiotemporal, Platonic, or "third realm" of thought, proposition, or bearer of truth (Frege 1956). In recent years, Trenton Merricks has delivered a similar idea: "I think that propositions are the primary bearers of truth and falsity. Moreover, I think that propositions represent things as being a certain way" (Merricks 2009, p. 207).

Secondly, others regard it as the referent the that-clause denotes or stands for, where the that-clause is often considered as a predicate modifier (Kratzer 2006), (Moltmann 2013); (Moulton 2009, 2015); (Elswyk 2019). Here is, for instance, what (Moltmann 2013, p. 123) writes: "that-clause is taken to stand for a proposition" (p. 123). Ikeya also writes that "we treat to PP [prepositional phrase], to VP [verb phrase] and that clause as constituting predicate modifiers" (Ikeya 1995, p. 155). In sum, it is widely accepted that that-clause comprises a full sentence that expresses a proposition.

And lastly, some look at it as the object of mental states such as thought, belief, hope, desire, fear, contemplation, perception, love, hate, mourning, etc. As a result, all those mental attitudes, states, or events are said to be propositional attitudes. In (Merricks 2009)'s words, "I shall use 'propositional attitude' to mean an attitude that is analyzed as-or reduced to-a relation between an agent and a proposition" (210). That is, a propositional attitude is a mental state that articulates a relation between a subject and a proposition. This philosophical practice of characterizing mental states as propositional attitudes starts with Bertrand Russell, who, in "Are beliefs, etc., irreducible facts?", writes:

> "Etc." covers understanding a proposition; it covers desiring, willing, any other attitude of that sort that you may think of that involves a proposition. It seems natural to say one believes a proposition and unnatural to say one desires a proposition, but as a matter of fact that is only a prejudice. What you believe and what you desire are of exactly the same

> nature. You may desire to get some sugar to-morrow and of course you may possibly believe that you will (Russell 1919, p. 50).

This Russellian attempt at providing a propositional analysis of mental states continued to dominate philosophy to the extent that it is widely believed that "fundamentally, all attitudes are propositional attitudes" (Grzankowski 2016a, p. 819). The following is a set of examples of philosophers who opt for the Russellian idea (Perry 1994; Stoljar 1996; Lemos 1994; Fodor 1978) and so on and so forth. Now, as it turns out, propositional attitudes must basically be analyzed as relations between a subject and a proposition. According to Grzankowski, "propositional attitudes are said to be mental or intentional relations holding between subjects and propositions" (Grzankowski 2016b, p 317). A theory of propositional attitudes should account for this relation.

So far, we have seen that such theories are dominating in philosophy. But what about non-propositional attitudes, if any? Recently, some philosophers, unlike propositionalists, have successfully, I believe, argued that propositionalism is at best imprecise and accordingly argued for non-propositionalism (Ben-Yami 1997; Brewer 2006; Crane 2001, 2009; Buchanan 2012; Grzankowski 2012, 2013, 2016a, b; Merricks 2009; Montague 2007; Thagard 2006). For instance, it has been argued that mental states like loving, belief, perception, liking, hoping, hating, fearing, and so on are attitudes but not propositional. Furthermore, some argue that mental states may be relational without being propositional. In other words, "there are intentional states relating subjects to propositions that are not propositional attitudes" (Grzankowski 2016b, p 317).

Here, by non-propositional attitudes, it was meant attitudes that "are [not] evaluable for truth, accuracy, satisfaction, or so on" (Grzankowski 2013, p. 1124). As expected, a considerable number of philosophers hold that concentrating on propositional attitudes amounts to neglecting the actually non-propositional attitudes. (Grzankowski 2012) tackles what he regards to be the main reasons behind this neglect.

One reason is that several philosophers have equated what is intentional with what is propositional (Stoljar 1996; Perry 1994). Analogously, others have considered propositions to be the objects of attitudes (McGrath and Frank 2018). That is to say; an attitude must be about an object that is a proposition. Second, some philosophers regard propositional attitudes as the most fundamental attitudes among the intentional ones, and all non-propositional attitudes are reducible to them. For example, (Bealer 1982, p. 225) writes:

> however, I have said nothing yet concerning the intentionality of intentional acts, i.e., their directedness or aboutness. How does that arise? The answer is that it arises from propositional objects, i.e., from the thought to which a person stands in the relation of thought, belief, judgement, etc.

Here non-propositional attitudes have not been defined away, yet they were considered as mediative, derivative, or not central. Bealer writes:

> nothing ever has been or ever will be (or ever has had or ever will have the capacity to be) conscious in any fashion-to have sense experiences, pleasures, pains, realizations, fears,

> wishes, and so forth. In such a situation, the envisaged etiological system would, intuitively, not have genuine intentional states (Bealer 1993, p. 115).

In this version of propositionalism, Bealer investigates a universe in which everything is either a fundamental propositional attitude or a derivative non-propositional attitude. Bealer, as a result, leaves aside those attitudes that are merely non-propositional and focuses on propositional attitudes.

It is with this in mind that we have to look at Bealer's view of intuition. That is, we have to do that in light of his notion of propositional and non-propositional attitudes. As we have seen, for him, intuition is a unique or sui generis, irreducible, natural propositional attitude. That is, it is the most fundamental, intentional, sui generis propositional attitude. And, all non-propositional attitudes are reducible to it. In accordance with such apprehension, intuition is conceived of as belonging to the category of fundamental propositional attitudes. What actually happened in this practice is that Bealer introduces the category of intentionally genuine propositional attitudes and then moves straight on to a discussion of intuition as one of them. Similarly, Sosa mentions that by intuition, he specifically means propositional intuition that is not derived from certain sources such as perception, memory, introspection, reasoning, etc. (Sosa 2006, p. 213). Also, we have seen that for Chudnoff, mental states like perceptual, intuitive, introspective, imaginative, and recollective experiences both represent the relevant proposition and make it seem as if the subject is aware of its truth-maker. As a result, Bealer, Sosa, and Chudnoff regard the state of intuition as propositional.

But what if there is a category of intentional but non-propositional attitudes? What if intuition belongs to it? To argue that such a category includes intuition is there is to argue that Bealer, Sosa, and Chudnoff are wrong about the category into which intuition belongs.

In what follows, I will adopt and defend a non-propositional analysis of mental states, arguing that it is applicable to intuition. To that end, I have to successfully argue that there is no good reason to characterize mental states as propositional attitudes.

The first reason why philosophers characterize mental states as propositional attitudes is the form of sentences that ascribes propositional attitude to a subject. Such a sentence takes "the form 'S Vs that p,' where V is the verb used to ascribe the mental state and p a proposition" (Ben-Yami 1997, p. 84). An example could be "Gettier believes that Smith does not know," where "believes" is V or the mental state verb and "Smith does not know" is P or proposition.

In other words, propositionalists found it rewarding to argue that mental states are propositional attitudes in virtue of 'that-clause', where 'that-clause' is understood as a term that refers to objects of mental states. As an example, one may think of Tyler Burge, who writes: "thoughts, beliefs, intentions, and so forth are typically specified in terms of subordinate sentential clauses, that-clauses, which may be judged true or false" (Burge 1979, p. 74). That is, that-clause refers to the proposition as objects of the given mental state. Here are some examples of it.

(1) John believes that **taxation is unfair**
(2) Jenny desires that **you deal with your problems.**
Nevertheless, it is not the case that every sentence takes the same form, even if it includes a verb used to ascribe the mental state. Consider the following examples.
(3) John wants you to come on time
(4) Jenny knows how to deal with her problem

By the same token, the reason why Bealer characterizes the mental state of intuition as a propositional attitude may be the form 'S Vs that p' that ascribes intuition, as a mental state, to a subject, as in 'Jackson intuits that our having all physical information about the subject won't help us to know everything about his color experience', 'Goldman intuits that Henry knows, or Goldman intuits that Henry does not know'. However, it does not seem to be the case that every intuition sentence takes the same form, even if it includes a verb used to ascribe the mental state of intuition, as in 'Jackson intuits the incompleteness of Mary's previous physical knowledge', 'Lehrer intuits the ignorance of Truetemp regarding what the temperature is when he truly says what it is,' and so on. As a result, if the reason for characterizing intuition as a propositional attitude is the form of a sentence, then it cannot be so characterized all the time.

However, to say that not every sentence takes the form of 'S Vs that p' does not entail that 'that-clause' is not a reporting or a referential term. Consider the following example: 'Kripke intuits that we do not refer to Schmidt when we use the name Gödel'. In this that-clause example, Kripke intends to tell us whom we do and do not refer to when we use the name Gödel: we do not refer to Schmidt but to Gödel himself when we use the name Gödel. But that doesn't mean that what was referred to by the 'that-clause' is a proposition. One could say that what was intuited was what the name Gödel referred to, which was expressed as a proposition when Kripke was lecturing, but when he intuited it, this was not necessarily a propositional intuition.

All that one may refer to by using a that-clause is what Ray Buchanan calls a kind of 'communicative intentions,' which, in turn, serve different purposes. The purpose might be informing, meeting social expectations, imagining, and the like. Consider the following example: 'BonJour intuits that Norman is epistemically unjustified in believing that the President is in New York City.' In uttering his intuition using that-clause in this example, all that BonJour may indeed intend to report is that his subject is not justified in believing what she does. However, "communicative intentions is not a proposition but rather a property of propositions … [which] is "associated" with many non-truth conditionally equivalent propositions" (Buchanan 2012, p. 17). If so, then there is a prima facie reason to reject the thesis that 'that-clause' refer to propositional entities.

However, many go in for characterizing mental states as propositional attitudes because they think that propositional attitudes are required to solve the long-term puzzle in the philosophy of language, namely, the puzzle of the empty name, i.e., a fictional proper name without referents (Orenstein 2003). Consider the following example: 'John believes that Pegasus has two wings.' How can Pegasus, which does not exist, have two wings? Propositionalists argue that taking mental states as propositional attitudes resolves the problem. The suggestion is that we have to consider the mental state of 'belief' to be in relation to the proposition 'Pegasus has two wings.' One can have a Russellian analysis imposed on this, removing the name "Pegasus" and replacing it with the descriptions that are attached to Pegasus.

To this, one can first object that this suggestion is not the case when it comes to examples such as: 'John loves Jenny', 'John likes Jenny', 'John hates Jenny', etc. In these examples, the mental state verbs are about 'Jenny'. That is, they are about an object, not propositions. They refer to Jenny without predicating anything about her. In the same sense, intuition is a non-propositional attitude since one can intuit an object without predicating anything about it. For example, 'John intuits four'. In this example, John intuits a number, the number four, and does not predicate anything more about it. The same considerations go for the example: 'Truetemp intuits the temperature.' In this example, the mental state verb 'intuit' is about the object and not a proposition with a single predication. In all examples discussed above, mental state verbs are not propositional, and they lack predication; yet that does not rule out intuiting, loving, liking, and hating from the relevant mental state verbs.

On the other hand, it is not clear how this suggestion makes the latter proposition, 'Pegasus has two wings', which involves a meaningless name, meaningful, after all. The meaningless name may make some sense if and only if we adopt the Russellian analysis. But why should we do that? Apart from that, Ben-Yami argues that:

> this approach would not work with states like, for instance, love: 'x loves y' cannot be construed as a relation between x and a proposition. Nevertheless, many people love, admire, despise, and so on, fictional characters. So some relations or attitudes to non-existent objects cannot be explained as actually being relations to propositions (1997, p. 87).

Analogously, the story goes for intuition. Take this example: 'Chalmers intuits that the realm of Zombies or Zombie itself (the subject of the story) is, indeed, conceivable.' It is not clear how the meaningless name 'Zombie' becomes meaningful just by following the aforesaid suggestion, i.e., taking the mental state verb 'intuit' as a relation to the proposition of 'the realm of Zombie or Zombie itself (the subject of the story) is, indeed, conceivable.' Nor can the mental state verb 'intuit', in accordance with Ben-Yami's view, be explained as related to the relevant proposition, even if many people do intuit fictional characters.

However, suppose that Ben-Yami is wrong and that mental state can be explained in terms of relation to the proposition. That won't make mental states propositional simply because, in such case, it will have the proposition as an object. Notice that in all the above discussion, it is admitted that mental state is intentional (i.e., it is about an object), yet it is not propositional. Here, I follow Grzankowski, who, by the word 'object,' understands, on the basis of the representational theory of mind, "a term in the language of thought that refers (or putatively refers if there is no referent) to o," where o = object (2016b, p. 318). This point will be clearer in the discussion below that focuses on differentiating between the two terms 'objectual' and 'propositional.'

Being intentional means that the mental state is directed to an object. And that object might be a proposition. But that might be enough to make a mental state an objectual attitude. Here I have to maintain that I am not following (Montague 2007; Forbes 2000)'s usage of the term 'objectual attitude.' They use 'objectual attitude' to mean attitude to an object other than proposition. But I use the very term to mean attitude to an object, even if that object is a proposition. But being objectual in a sense defined above does not make a mental state propositional. That is, it won't be enough to make it sensitive to the truth conditions under which the given proposition is true or false. An example can be 'Jenny fears the proposition that Superman exists.' The mental state verb 'fear' is intentional; it is directed to an object, i.e., 'the proposition that Superman exists.' Furthermore, being directed to an object, even if that object is a proposition, makes the relevant mental state verb objectual but not propositional. Similar considerations can be applied to intuition, as in 'Gettier intuits the proposition that Smith does not know', 'Cohen intuits the proposition that the subject does not know he will lose,' 'Cohen intuits the proposition that the subject knows he will lose,' and so on. Although the mental state verb 'intuit' is intentional- it is directed to an object that might be a proposition- and accordingly, objectual, it is not propositional simply because it is not sensitive to the truth conditions of the given proposition.

A mental state with an object or a proposition as an object is similar to empty names, which are "representations that refer to things" that might not exist. As Grzankowski says:

> one might fear Pegasus even though Pegasus doesn't exist and in fearing Superman one needn't fear Kent. It is hard to see how to make sense of these facts without appeal to representations (Grzankowski 2016b).

These representations seem to be de dicto: they are related to the form of expression itself (e.g., Superman, Pegasus, etc.) rather than any property of its referent (e.g., being strong). Consider the following examples. 'Jenny fears Zombie' and 'Searle intuits the proposition that the computer (the protagonist of the story) lacks cognitive states.' 'Zombie' and 'the proposition that the computer (the protagonist of the story) lacks cognitive states' are representations that refer to objects. These objects here are non-existent entities in the case of 'Zombie' and proposition as an object in the case of 'the proposition that the computer (the protagonist of the story) lacks cognitive states.'

Propositional object is not enough to make a mental state propositional. However, for a mental state to be propositional, it must be a representation that has a propositional content (Grzankowski 2016b).[20] All this leads us to (Grzankowski 2013)'s idea that non-propositional attitudes are intentional mental states. They have aboutness, i.e., they are about things, objects, propositions, etc. This aboutness is not

[20] Sosa won't have any problem with the notion of the propositional content of intuition. Yet, the problem is with his idea that this propositional content is expressed by concepts, and we are attracted to assent to this content based only on consciously understanding those concepts. I have already objected to his idea of both understanding and concepts.

mediated by subjects' relations to propositions regarding those things but by virtue of other things such as mode of presentation, relations to non-existent entities, etc.

If all I have argued for in this section is true, then mental states, intuition included, are attitudes but not propositional. But suppose I am right: Intuition is non-propositional. So what? Why does all this matter at all? For at least one reason: whether intuition is propositional or non-propositional is relevant to the question about the nature of intuition which inherits interest from the question about the epistemic status of intuition since how we answer the former affects how we answer the latter. Intuition, some have thought, is propositional. Not so, if what I have argued for is correct. Realizing that there is a prima facie case for intuition being non-propositional teaches us that it cannot be judged as true or false. Thus, the current debate on whether intuition is true or false loses its ground of legitimacy.

But the one, who seeks a correct account of the nature of the mental state of intuition, is supposed to wonder whether it possesses a phenomenology. I have argued for the intentional but not the propositional character of intuition. However, I do not want to presuppose that such a mental state is phenomenal or it has phenomenal character. Accordingly, in the next section, my main objective will be to question whether such a phenomenology exists.

2.3.3 Non-phenomenal Intuition

We have seen that, following Husserl and Bealer, Chudnoff argues for the phenomenality of intuition. The term 'phenomenology' refers to a metaphysical assumption of what the world is like and to an epistemological notion of how the world can be known (Berrios 1989, p. 425). In the spirit of this view, Chudnoff develops his definition of intuition's phenomenality as follows: in having an intuition that p, it is like to be that p or it seems to you that p. However, intuition is underlain by other mental states that I have successfully, I believe, shown as non-propositional attitudes such as imagination, thought, and so on. Also, for him, their phenomenal character constitutes that of intuition. He explains the special phenomenology such mental states possess as follows:

> there is something it is like for you before understanding, intuiting, seeing, or reacting. There is something it is like at the moment of understanding, intuiting, seeing, or reacting. And what it is like before being in these mental states is different from what it is like while being in these mental states (Chudnoff 2015, p. 2).

Bealer argues that "for you to have an intuition that A is just for it to seem to you that A" (Bealer 1996b, p. 5). Even Sosa, as we have seen, did not argue against the idea of seeming. What he denies is the claim that seeming does serve as a regress-stopper. However, in classifying mental states as phenomenal, Bealer, Sosa, and Chudnoff do not stand alone. In particular, this position was also expressed by David Braddon-Mitchell and Frank Jackson. They draw a distinction between sensory and cognitive states, holding that sensory states have a distinctive phenomenal character, but cognitive states do not. Here is a helpful and representative example of this standpoint:

> bodily sensations and perceptual experiences are prime examples of states for which there is something it is like to be in them. They have a phenomenal feel, a phenomenology, or, in a term sometimes used in psychology, are raw feels. Cognitive states are prime examples of states for which there is not something it is like to be in them, of states that lack a phenomenology (Braddon-Mitchell and Jackson 2007, p. 129).

That is to say that if someone stabs me in the stomach, then there will be something it is like for me to have stabbing pain which feels different from a bone fracture. But if I intuit that knowledge is justified true belief, then there won't be something it is like for me to have this intuition. Nor will this intuition be different from the intuition that Aramaic is a living language.[21] Does that mean that we can exclude cognitive states from the list of phenomenal states? The answer to this question comes from Goldman in the negative. The central task of Goldman's "The Psychology of Folk Psychology" is to instigate a phenomenological approach to mental states, both sensory and cognitive. To that end, he cites Jackson's anti-physicalism example of the Mary case and suggests that a similar argument for attitudes is possible. He writes:

> just as someone deprived of any experience of colors would learn new things upon being exposed to them, namely, what it feels like to see red, green, and so forth, so (I submit) someone who had never experienced certain propositional attitudes, for example, doubt or disappointment, would learn new things on first undergoing these experiences. There is "something it is like" to have these attitudes, just as much as there is "something it is like" to see red (Goldman 1993, p. 24).

To understand the view, we have to flesh it out a bit further (albeit at the risk of simplifying somewhat). There is something it is like to have a stabbing pain, just as much as there is something it is like to believe that knowledge is justified true belief. Accordingly, both sensory and cognitive or intellectual states have a phenomenal character. In the same vein, David Pitt defends the view that mental states are conscious thoughts that have distinct phenomenal properties. He writes: "there's something it's like consciously to think that p, which is distinct from what it's like consciously to think that q" (Pitt 2004, p. 1). Simply put, what it is like to be in a mental state is different from what it is like to be in another. For instance, what it is like consciously to think that Modi is the prime minister of India is distinct from what it is like consciously to think that Aristotle is a great philosopher. In each case, there is something special it is like to be in it.

Furthermore, other philosophers did not only view mental states from the phenomenological angle, but they also argued that mental states are best understood through their distinct phenomenal character. Here is an instance of this perspective: "if you don't know what it's like to feel sad, you can't understand what it is to feel sad. And if you can't understand what it is to feel sad, you can't regard something

---

[21] Aramaic is the ancient Semitic language of the Arameans in Syria. Aramaic was the mother language of Jesus, and it is still spoken nowadays as a first language by many communities of Syria.

as feeling sad" (Robbins and Jack 2006, p. 70). That is to say, to know the nature of mental states well, one has to consider them as phenomenal states. All in all, mental states have intrinsic phenomenal character.

This raises an important question. Does the above picture justify the belief that classifies mental states as phenomenal? The answer Nichols and Stich construe is a simple no. They straightforwardly put the phenomenal properties of the mental state into question: "believing that 17 is a prime number doesn't feel any different from believing that 19 is a prime number. Indeed, as best we can tell, neither of these states has any distinctive qualitative properties" (Nichols and Stich 2003, p. 196). In a similar vein, Ludwig describes the view that there is a distinctive phenomenology for every type of judgment, 'a defect'. He writes:

> a defect of the phenomenological classification is that it is unclear that every judgment which expresses solely conceptual competence has a distinctive phenomenology. The phenomenology of intellectual intuitions cannot simply be that something seems to us to be so. Things can seem to us to be a certain way though we are not having an intuition because the seeming has the wrong sort of explanation (Ludwig 2010, p. 433).

In short, Ludwig defends the assumption that intuition is accompanied by a certain phenomenology as incorrect. In what follows, I will try to be consistent with this view. An option for me to do so will be to adopt a view according to which there is hope to show in a convincing manner that mental states, including intuition, are non-phenomenal. If (Chalmers 1996) does not find the phenomenology of mental states worth fighting for, my view is to argue that their non-phenomenality is worth fighting for. My strategy to achieve this aim lies in the argument from case studies.

The Argument from the Case Studies The examples of thought experiments considered in the first chapter provide evidence that no mention of a distinctive phenomenology is required to explain the emergence of the relevant intuitions. As an illustration, consider again the Chinese Room case, where Searle imagined himself as ignorant of Chinese and locked in a room with Chinese writing, script, symbols, and Chinese rules and with instructions written in English. Yet, behaving like an intelligent computer, his performance in Chinese was just as any native Chinese speaker's. Searle's Chinese Room case depends on the intuition that the computer lacks cognitive states. Searle presented his case. Nothing about it strikes us as *seeming*, in some special phenomenological sense, to be true. So, we can wonder whether this kind of 'seem true' feeling exists at all. Searle presented his case in a way that clearly doesn't attempt to appeal to any such feeling. I am not aware of any contribution to the discussion of this thought experiment that appeals to a special phenomenology when the intuition about this case arises. Nor did Searle himself give a hint that he had such a special feeling when the intuition about this case was contemplated.

Consider again Lehrer's Truetemp case in which Lehrer invites us to imagine a person called Mr. Truetemp who underwent a brain operation. The surgeon implanted in Truetemp's head a device that enables him to think truly about temperature. Lehrer's intuition is that Truetemp does not know that the temperature is

104 degrees when he correctly thinks it to be so. At no point in this thought experiment does the special phenomenology of intuition play any role. The author is unaware of any such special phenomenology, and so he certainly did not mention either explicitly or implicitly that such a special feeling accompanied his intuition.

In short, neither Searle nor Lehrer thinks that he required a distinctive phenomenology to come up with his intuition. The same approach goes for the rest of the epistemic thought experiments, especially those which were considered in the first chapter. That being so, I conclude that the presence of a special phenomenology that accompanies intuition is not present in the original texts where cases were first presented. Thus, philosophers do not appeal to phenomenology as such when intuiting.

It may be thought that in Frank Jackson's "Robot" Mary case, clearly, Mary has a new phenomenological experience of red or whatever color when she sees all the flowers in the garden. That may well be true, but this does not show that the *intuition* that Mary had a new experience in *itself* has a special phenomenology to it. Did Jackson say, "I think I feel something special about the way my intuition feels about the way Mary felt after seeing red?" No.

The same goes for Chalmers' case of Zombies. Whether this case shows that physicalism is false or not, Chalmers has never claimed that the intuition he has had some special phenomenological feel about it. Did Chalmers phenomenologically feel that physicalism is wrong? What does such a feeling *feel* like? As can be seen from the debates all philosophers carry about these cases; it is clear that no philosopher ever says that she is winning because of the phenomenological way her arguments feel.

We have come to the conclusion that thought experiments provide evidence that their authors do not appeal to any phenomenological property along with which they arrived at the relevant intuitions. On the contrary, philosophers provide some thought experiments as evidence that mental states, in general, and intuition, in particular, have no phenomenal properties. But if not by phenomenology, what individuates intuition?

2.4 The Individuation of Intuition

Intuition is said to be individuated in terms of phenomenal character. This suggestion was already rejected. Some want to individuate intuition in terms of other mental states, namely, emotions. Keeping in mind the argument provided in the previous section that intuition comes without a special feeling or what it feels like, this view is no more acceptable. The story goes for the view that individuates intuition in terms of its being a propositional attitude. Since I have argued that intuition is non-propositional, this perspective is no longer satisfactory. And others, as we have seen in the case of the three intuition-theorists' accounts considered in this chapter, individualize intuition in terms of understanding. On the understanding of individualization, intuition is taken to be composed of a priori concepts. But we have shown how the view that considers concepts to be pure a priori in nature is misguided. However, one can contemplate a new account of intuition individualization.

The suggestion is that intuition can be individuated in terms of its origin, namely, thought experiment. Intuition is supposed to have a lot in common with the thought experiment. For example, it is true that the statue may be distinct, in terms of form, color, size, etc., from the clay out of which it is constituted. Nevertheless, both the statue and clay are made exactly from the same feedstock. In the same manner, intuition emerges from the thought experiment, and accordingly, it shares at least some of the characteristics of the thought experiment already identified. In view of that, we may talk about the intuition state as directed at an object, where this directedness is taken in its intentional terms. When I intuit, my intuition is about an object. Consider Jackson's intuition that our physical knowledge is incomplete. It is obvious that the object at which the intuition is directed is knowledge. The same story goes for other intuitions of thought experiments we have discussed in the previous chapter. In this sense, intuition was considered to be an intentional or objectual mental state but not propositional.

This leads us to the informativeness of intuition. That is, by virtue of intuition, we learn something new about its object. Intuition can be constructively informative: it aims as a part of a whole at an act of establishing evidence for a theory, as in Plato's intuition that knowledge is true belief with an account.[22] It can also be destructively informative; it aims -as a part of a whole- at an act of destroying an old theory and generates no new one as in Gettier's intuition that knowledge is not justified true belief. Lastly, it may be Platonic: it aims -as a part of a whole- at both an act of destroying an old theory and an act of establishing evidence for a new theory at the same time. For instance, Galileo's intuition that all objects fall alike. This intuition is platonic because it refutes Aristotle's view that weighty objects fall quicker than light ones and at once establishes evidence for an alternative theory. At any rate, by virtue of intuition, we learn something new about its object.

And, if intuition has constructive, destructive, or Platonic informativeness, then this illustrates a significant feature of it, namely, its independence from the assertions regarding its object. Our intuition is not sacred and so deserving veneration; nor is others'. In fact, our intuitions conflict from one brain to another, as in Ayer's and Gettier's intuitions regarding knowledge. Variation in intuition is widely-discussed but ill-understood. The technique of thought experiment, whether constructive, destructive, or Platonic, brings us to new knowledge: the intuition of such thought experiment enjoys its powers and limits as an intuition of exceptional cases. What is new and exceptional is, I think, controversial. This does not mean that we do not often idealize our intuitions, thinking about them as perfect.

Consider BonJour's intuition that Norman's belief that the President is in New York City is mere luck. If BonJour did not look at this intuition as ideal, he would not have considered it to be a wrong instance of knowledge. Of course, not every intuition is controversial; some are universally accepted. To differentiate between these two kinds of intuitions, we need a standard which I will set in the next chapter.

[22] By the term 'whole,' I mean thought experiment as an argument. So conceived, there are some great ways in the teachings of intuition and thought experiments that reinforce each other. I will come back to this idea later in the next chapter.

2.5 Concluding Remarks

In this chapter, I have developed an account of what intuition is. This account maintains that intuition is intentional or objectual, but it is a non-propositional, non-phenomenal mental state which can be individuated by its origin. And such intuition transcends the traditional a priori a posteriori distinction toward a multi-dimensional space. This stipulation is likely to raise a further question we are now supposed to be poised to address: what epistemic weight does intuition carry? It is this question that I will address in the next chapter.

Chapter 3
The Epistemic Status of Intuitions

3.1 Introduction

The main goal of this chapter is to assess the various positions and arguments that epistemologists have given for and against the epistemic status of intuitions. I plan to do that in light of the conception of the nature of intuitions developed and defended in the last two chapters. It must be made clear that I am taking account of intuitions developed at the end of chapter one, transferring through chapter two to be the correct general account of the nature of epistemic intuitions, and the positions and arguments given in this chapter will be assessed in light of that account. Moreover, this chapter will be concerned with the question of whether epistemic intuitions have evidentiary status.

There is a wide range of questions one could pose about the epistemic role of intuition. I focus on the basic issue of whether intuitions have epistemic statuses or whether they are instead outside the scope of epistemic evaluability. More specifically, this chapter will be concerned with the following question: when, if ever, do our intuitions qualify as evidence, and in virtue of what do they do so?[1] I must also make clear that the main concern for the epistemic status of intuitions will be whether they can play the role of epistemic justifications in the form of arguments. In other words, the main consideration for this chapter will be whether intuition is apt for such a role. This is what will be meant by the expression 'epistemic status' of intuition.

[1] I leave it open whether philosophers rely on intuitions as evidence or as sources of evidence. In doing so, I follow Bealer, who writes that intuitions are evidence (see, e.g., Bealer 1996a, p. 2). However, in other passages, he considers intuitions as a source of evidence (see Bealer 1992). Nonetheless, it is difficult to run the debate about the epistemic status of intuitions in complete isolation from the endorsement of a particular account of evidence. Even though I stay neutral on that issue, I will talk only of 'intuitions as evidence,' by which I will mean 'intuitions as evidence or source(s) of evidence.'

M. Hamdo, *Epistemic Thought Experiments and Intuitions*, Philosophical Studies Series 150, https://doi.org/10.1007/978-3-031-33480-1_3

This assumption will concern many epistemologists. I will start this chapter by considering the main positions on the epistemic status of intuitions. These positions will be broken down into three main camps: reliablists, skeptics, and perspectival relativists. Reliabilism concerning the epistemic status of intuitions maintains that intuitions are essential to epistemology and that one's having an intuition that p provides epistemic support for one's being justified in believing that p or for one to know that p. Such a position is mainly maintained by Bealer, Sosa, Chudnoff, and others. While there are slight differences between the various advocates of reliabilism, reliabilism about intuitions remains the main thesis that intuition evidentially qualifies as a reliable source of knowledge. However, in addition to expository prologues to these three views, I will be trying, at least, to undermine Bealer's reliablist form of modal intuition, Sosa's competence-based view, and Chudnoff's phenomenal dogmatism.

On the other hand, skepticism (the anti-*intuition* camp) concerning the epistemic status of intuitions maintains that intuitions are epistemologically useless. There is something either about variation in intuitions or the lack of ability to calibrate our intuitions that leads the skeptics to the conclusion that intuitions have no epistemic weight and should be excluded from epistemology. It must be noted that most of the skeptics' arguments for the unreliability of intuitions are due to their concern with the nature of intuitions as cashed by reliablists. Most of the skeptics' worries about the epistemic status of intuitions are worries about the epistemic status of a priori intuitions. Skeptics have not concerned themselves with an account of intuitions as was developed in chapter two. This position has become increasingly popular in recent years. In this regard, I will assess two arguments in support of skepticism about the epistemic status of intuitions: the variation argument and the calibration argument. Then, I will show that both of these arguments are problematic.

Lastly, perspectival relativism (the middle path camp) holds that intuitions ought to and do count as evidence in support of some claim (contra the skeptics), but we need to rethink how this happens in light of considerations derived from several perspectives and disciplines in the process of wide reflective equilibrium and how much support intuitions can give (contra the reliablists). Simply put, perspectival relativism concerning the epistemic status of intuitions rejects the absolute epistemic status of intuitions but does not believe epistemic intuitions to be false. But I render this position doubtful.

Having undermined the main positions on the epistemic status of intuitions, I will try to articulate what seems to me to be the right position that any adequate account must respect: intuition alone has no epistemic status. That is to say, intuition combined with other pieces makes up a whole. One part of the whole is the thought experiment that generated intuition. If so, the question of epistemic status does apply to the whole and not the intuition alone, and if it is true that the epistemic argument is different from the logical argument, then they both also must be appraised differently both in the sense of the standards of appraising as well as the appraisal words. Here is the plan for the chapter.

In Sect. 3.2, I will discuss reliabilism about intuition and raise some skeptical worries about it. Similarly, in Sect. 3.3, I shall address the challenges of experimentalism. After that, in Sect. 3.4, I will consider perspectival relativism about intuition and articulate some criticism of it. In Sect. 3.5, I will be stating my desiderata for the epistemic status of intuition. Finally, in Sect. 3.6, I will conclude, wrapping up the various threads of discussion.

3.2 Reliabilism About Intuition

By reliabilism about intuition, I mean the view according to which intuitions are to be regarded as a reliable evidential source of knowledge. To see more precisely how reliabilism about intuition proceeds, we must begin with its characteristics shown by reliablists themselves. Keeping in mind at this point that I am in no way committing myself to one particular framework here, I will focus on the main paradigmatic forms of this view in what follows.[2]

3.2.1 Bealer on Modal Reliabilism

Initially, Bealer's account seems enticing. He explains how intuition qualifies as evidence by introducing what he calls 'Standard Justificatory Procedure,' according to which, he describes, just like many other items, such as experiences, observations, phenomenal experiences, or testimony; in our standard epistemic practice, intuitions are to be treated as evidence or reason to justify various conclusions, beliefs, or theories (Bealer 1992, P 100; 1996a, p 122; b, p 3; 1998a, p 204–7; 1999a, p 30; b, p 246; 2002, P 73; 2004, P 13; 2008, p. 191). By 'evidence', he understands the content of the intuition intuited by an aware individual. While intuiting that Mary will learn something new about the colors in the world after she gets released from her black-and-white room, Jackson is aware that he is intuiting that. If so, philosophers consciously use intuition as evidence in their argumentative theorizing. This usage, according to Bealer, takes "the following idealization: canvassing intuitions; subjecting those intuitions to dialectical critique; constructing theories that systematize the surviving intuitions; testing those theories against further intuitions; and so on until equilibrium [among intuitions] is approached" (Bealer 1998a, p 205). As it is described by Bealer, 'Standard Justificatory Procedure' is just a different name of the method for reflective equilibrium.

Once the method of Standard Justificatory Procedure takes place, his problem with naturalism or empiricism becomes clear. According to him, his opponents, naturalists, are inconsistent. On the one side, they hold that the use of intuitions as evidence in philosophy is what ultimately plagues philosophy. Evidence, for

[2] Bealer, Sosa and Chudnoff are obviously not the only defenders of the reliability of intuitions. I particularly discuss their arguments only because they have presented substantive views of such types of defense in very recent literature.

naturalists or empiricists, comprises one's experiences and observations, excluding intuition.[3] On the other side, they do allow intuitions to act as evidence and thereby end up in an epistemically self-defeating position. They do rely on intuitions regarding, e.g., twin-earth scenarios as evidence for the truth of scientific essentialism. It is, after all, philosophers' intuitions about twin-earth-style scenarios which led us to conclude that terms like "water" are semantically like indexicals. However, for him, empiricists, when challenging the evidential status of intuition, do not understand by the term 'intuition' what he does, namely, rational intuition that expresses itself as necessary.

To reply to empiricists, however, he offers two arguments. The first is 'The Argument from Evidence' or a restricted form of reliabilism, according to which it is "only ... basic sources of evidence be reliable." And what is that basic source of evidence? He defends the thesis that "something is a basic source of evidence iff it has an appropriate kind of reliable tie to the truth" (Bealer 1998a, p. 215). This reliable tie to the truth cannot be contingent because if it is contingent, some experiences (e.g., accidentally-reliable guesses) will be counted as basic evidence, but they should not be so considered. So, contingent reliabilism is false.

The most convincing theory left is 'modal reliabilism,' according to which "something counts as a basic source iff there is an appropriate kind of strong modal tie between its deliverances and the truth" (Bealer 1998a, p. 216). But what kind of modal tie is appropriate? Bealer suggests "the weakest modal tie that lets in the right sources and excludes the wrong ones" (Bealer 1998a, p. 218). According to this weakest modal tie, "a candidate source is basic iff for cognitive conditions of some suitably high quality, necessarily, if someone in those cognitive conditions were to process theoretically the deliverances of the candidate source, the resulting theory would provide a correct assessment as to the truth or falsity of most of those deliverances" (Bealer 1998a, p. 219). So here, not every but most of the deliverances are required to be true. "Intuition does have that sort of modal tie; hence, intuition is a basic source of evidence" (Bealer 1998a, p. 219). If so, Bealer's 'Autonomy of Philosophy' thesis holds. That is, philosophy can answer its central questions without getting help from science. But why should intuition have such a tie to the truth?

To answer this question, he introduces the second anti-empirical argument: 'The Argument from Concepts'.[4] As mentioned earlier, by 'evidence,' he understands the content of the intuition, the conceptual content of it. So, in order to track the truth of intuition as evidence in the sense of aforesaid weakest modal reliabilism, we have

---

[3] Nevertheless, he states that naturalists themselves use intuition. Accordingly, the Standard Justificatory Procedure is practically followed by naturalists who then, if they want to be consistent, have little room to legitimate their beliefs that are supposed to be relied on experience and observation and not on intuition. If so, if naturalism relies on intuition, naturalism is a self-defeating thesis.

[4] Bealer's account of concept possession is one of impressive scope. I am not discussing it here in detail. Nor am I going to evaluate it. All that the context of our research requires us to know about it is that it is an account of a priori knowledge and it is an account of modalities (possibility, necessity, contingency).

to investigate the question of how we possess concepts. According to Bealer, one can possess concepts in two ways. The first is that one can possess concepts in a minimal, weak, indeterminate, or incomplete sense, as in Burge's arthritis example (Bealer 1998a, b, 2004). The second is that one can possess concepts in a complete, determinate, full, or strong sense, as in the multigon and chromic examples (Bealer 1987a, 1998a, b, c). Rational intuition is grounded in the second way.

It lies in our abilities to grasp and apply its concepts in an a priori stable fashion with a high level of cognitive conditions. We are able to do so, as in the Multigon and Chromic examples and most of the central concepts of philosophy. So, intuition has the above-mentioned tie to the truth (i.e., rational or a priori intuitions are necessarily true). If so, the 'The Authority of Philosophy' thesis holds. That is, philosophical answers to questions are stronger and greater than scientific ones (Bealer 1998a). But this amounts to a clash with scientific essentialism, the view advocated mainly by Saul Kripke and Hilary Putnam, according to which some necessities such as water = H20 "require empirical investigation, making them essentially a posteriori" (Bealer 1998d, p. 30).[5] That is, "water cannot be decided a priori by our modal intuitions; empirical science is required (Bealer 1994, p. 186).

To solve this problem, he distinguishes between local scientific essentialism, which includes naturalistic concepts (e.g., water, heat, etc.) and global scientific essentialism, which includes content concepts that "are intended to include the concepts in terms of which (most of) the central questions of philosophy are formulated" (Bealer 1987a, p. 295). For him, the attack from scientific essentialism comes under the category of local scientific essentialism, which is excluded from philosophy. The relevant issue, then, is what saves the boundary between intuition and empirical science (Bealer 1987b). For him, it is saved by the two theses of the autonomy and authority of philosophy.[6]

The proposal is that scientific essentialism is not merely applicable to the central concepts in philosophy. Unlike concepts such as water, philosophical terminologies have the characteristic of being semantically stable. If an expression is semantically stable, it follows that the contingent factors of the competent utterer's milieu are not able to take part in constituting its meaning. That is the reason why a concept like knowledge is not a twin-earthable term. However, (Bealer 1996a) admits that there are some philosophical terms which are semantically unstable such as time, space, cause, and probability. Nevertheless, he proposes that these terms have generic uses which are semantically stable, as evidenced by phrases like 'Euclidian space is a

[5] Scientific essentialism is the thesis "that: (1) each scientific kind is associated with the same set of properties in every possible world; and (2) every individual member of a scientific kind belongs to that kind in every possible world in which it exists" (Khalidi 2009).

[6] It is worth mentioning that Hans-Johann Glock (2017) defends the idea of the autonomy of philosophy. According to him, when philosophy deals with issues that also interest science, philosophy is noncontinuous with science. That is, philosophers should not indulge themselves in the issues that also interest science. In cases as such, "scientists should be duly grateful rather than hostile" (100). But Glock does not defend the idea of the authority of philosophy over science. That is, when philosophy deals with issues that also interest science, philosophy is less a part of science, and it has no epistemic priority over science.

possible kind of space'. It is these generic uses which underlie the general philosophy of e.g., space and time.

Having spent some time exploring Bealer's account of the epistemic status of intuitions, I now proceed to raise difficulties with the proposal. The move just considered is, I think, open to serious question. The list of semantically unstable philosophical terms is possibly much larger than Bealer imagines. For example, the thesis that the term "knowledge" denotes semantic instability can be easily defended: it is Plato's supposed conception of knowledge as a justified true belief that Gettier subjected to criticism. Similar positions with regard to terms like intuition have been held by, e.g., experimentalists and reliablists.

Bealer's account deliberately models intuitions on modal intuitions: our intuitions are the foundations for our possible and necessary beliefs. But they are not. Take Gettier's Smith-Jones case, for example. It might not be straightforwardly necessary that Smith does not know that the one who will get the job has ten coins in his pocket. Gettier might partly or completely make up his scenario in a different way where the intuition that Smith does know is acceptable. Nor does reflection on Truetemp's case make it simply seem to you that necessarily Mr. Truetemp does not know what the temperature is when he truly says what it is. Philosophers have very different intuitions regarding whether Henry knows that there is a barn in front of him in the Fake Barn cases.

I take it that Lehrer's scenario of Truetemp can be argued to be incomplete. Having received a new perceptual faculty, Truetemp must also be given many neural circuitries and cognitive abilities through a long process of adaptation and development. Lehrer's scenario and the intuition that arises from it can be otherwise. To my mind, the reason behind this mistake of Bealer is based on the alleged intuition-perception analogy on a phenomenal basis. That is, just as with seeing an apple, the proposition 'I see an apple' seems to be necessarily true. The same story goes for intuition. I have already argued against the phenomenal character of intuition.

The thesis that the clash with scientific essentialism can be solved by merely classifying the scientific essentialism attack under the category of local scientific essentialism, whose concepts are not central to philosophy, is imprecise. It is true that scientific essentialists discussed natural kind concepts. For example, in "Meaning and Reference" (1973) and "The Meaning of 'Meaning'" (1975), Putnam argues for his thesis of semantic externalism, according to which the meaning of a word cannot be determined only by psychological terms, such as beliefs, images, etc. Similarly, in *Naming and Necessity*, Kripke discussed the same kind of concepts, such as gold, H_2O, etc. But scientific essentialists also discussed content concepts that are, according to Bealer himself, central to philosophy. For example, (Kripke 1980), along with the problem of natural kind terms, deals with the problem of names, intentionality, a priori, a posteriori, necessity, contingency, truth, essential properties, identity, etc. Unlike the traditional picture that holds that everything necessary is a priori and everything contingent is a posteriori, Kripke argues that there are both necessary a posteriori propositions and contingent a priori ones.

Regarding necessary a posteriori propositions, Kripke presents many examples. One of them is the statement that asserts the identity between Hesperus and

Phosphorus. He argues that we empirically know that Hesperus is Phosphorus. We need empirical evidence to determine that both names have the same reference. Then the statement is knowable a posteriori. Nevertheless, the moment we get this evidence, the statement that Hesperus is Phosphorus becomes necessarily true. This is the case because as soon as we get this empirical evidence, we will get to know that both names Hesperus and Phosphorus are rigid designators: they refer to the same object in every possible world where Hesperus and Phosphorus exist. Thus, we should accept that there are necessary a posteriori truths.

Regarding contingent a priori propositions, Kripke also presents many examples. One of them is the proposition "the length of stick S at time t_0 is one meter." He argues that this proposition is contingent a priori. For the one who is fixing the reference of "one meter", this proposition is a priori. To fix that, one just has to stipulate that the length "one meter" "is the length of stick S at time t_0". In other words, it is not a matter of learning or investigation. It is just a matter of reflection, which is enough to make one realize that the stick is one meter. But it is contingent because it is possible that the length of the stick was not one meter but because stresses and strains were less or more than one meter.

Given the fact that Kripke thinks that these examples of necessary a posteriori and contingent a priori are intuitively correct, two things are true. First, that scientific essentialism does not include only naturalistic concepts, but it does include content concepts that are, as Bealer himself says, central to philosophy. Second, if Kripke is right, then even if there are necessary truths, it would not follow that there can be only a priori knowledge of the general modal status of necessary intuitions. The crucial point is that if any of these two options is correct, then it would appear that Bealer ought to admit that it is not even obvious that his arguments for the reliability of intuition succeed. For that reason, Ernest Sosa has offered a somewhat different account of the epistemic status of intuition.

3.2.2 *Sosa on the Competence-Based View*

Ernest Sosa has offered an account of the epistemic status of intuition which disagrees with Bealer's idea of what gives intuition the status of being a basic source of evidence. He argues that Bealer discusses two possibilities that show why intuition is a basic source of evidence. One is contingent reliabilism, according to which intuition is a basic source of evidence because it has a contingent tie to the truth. Another is modal reliabilism, according to which intuition is a basic source of evidence because it has a modal tie to the truth. As construed by Sosa, Bealer uses five examples to show the contingency of contingent reliabilism to reject it, shifting to modal reliabilism. Examples include the case of the subject who is reliable, on the basis of telepathic manipulation or through hardwiring, in believing some contingent and necessary propositions. Sosa holds that the subject of any of these examples does not have only contingent but necessarily true belief. Hence, he criticizes this shift saying that the examples against contingent reliabilism are as effective as those against modal reliabilism. As a result, intuition is hardly a basic source of evidence.

What is more, against empiricism, Bealer defends intuition as a valid source of evidence by appealing to the standard procedure. The problem with this proposal, according to Sosa, is that such an appeal is

> not to the standardness of that procedure, but rather to the substantive considerations appeal to which is allowed by that procedure which is in fact standard among us. If this is the strategy, however, then the problem we face is one of relative emptiness, of insufficient content. If the reference via standardness is entirely accidental … then, until the substantive character of the procedures in question is revealed, we cannot assess the large claims made in their behalf (Sosa 1996, p. 161).

Another view Sosa rejects is the introspective view of intuition, according to which intuition is true because it gives us a direct insight into the abstract. Contrary to that, Sosa holds that if we accept the introspective view as true, then we justify "the epistemology of fallacious reasoning," as in one's affirming the consequent in the form "that, necessarily, if q, and p→ q, then p."[7] The word "necessarily" indicates that the intuition is clear and distinct and accordingly true, but it is not so (Sosa 2007a, p.57). For that reason, he comes to the conclusion that this model is inadequate for the epistemic worth of intuition. But what denies justification to the fallacious reasoning?

Sosa's own favored view of the epistemic role of intuition is the competence-based view which does not only make intuition rational, as we have seen in the last chapter, but it also makes it evidential.[8] In other words, intuition can attract someone if and only if: (1) it was based only on understanding, and (2) it was derived from competence which gives the understanding both its rationality and evidentiality.[9] On this view, intuition is evidentially true; because it is virtuous (i.e. it is derived from this virtue or competence). In addition, fallacious intuitions are counted to be one's "performance errors … by contrast with deliverances of a competence." That is, intuitions can be fallible as performance errors due to one's carelessness, inattentiveness, etc. Accordingly, "what denies justification to the fallacious reasoner might just be his carelessness or inattention or blundering haste." Sosa then does not deny that intuition may be false, but then again, they are reliable or epistemically justified because they are competent. They are reliable though they may sometimes

[7] Sosa is interested in modally true intuitions. If this is true, and if his account is to be a general theory of the truths, why does it restrict itself to tell the true from the false in the domain of the modally strong – the necessarily – true or false?

[8] He sees here an opportunity for his signature virtue-based account of justification. He distinguishes between Basis-Dependent Foundational Justification and Virtue Foundational Justification. The former is foundational justification that derives from the justified belief's being based on a psychological state of the subject, one that is beyond issues of justification; the latter is foundational justification that derives essentially from the justified propositional attitude manifesting an epistemic competence.

[9] But there is a standard worry about Sosa's account; namely, it makes it hard to know whether someone's intuitive judgment that p is or is not rational without first knowing whether this particular judgment is a manifestation of his capacity to reliably tell the true from the false.

be false due to inattentiveness or the like. The same story does not go for the perceptual and introspective models where false intuitions are not counted as reliable because "error is not just due to carelessness, moreover, nor to inattention or the like," but it is relative to … [the] system itself, simply because "the attraction is a deliverance of … [the] system itself" (Sosa 2007a, p.59).

If fallacious intuitions are so because of the subject's carelessness, inattentiveness, etc., then they are correctable by paying more care, attention, etc. At any rate, Sosa does not suggest that. He suggests that they are correctable by reflective equilibrium. In (Sosa 2009), the main aim is to save reflective equilibrium from Stephen Stich's attack. By 'reflective equilibrium,' he understands that we reach a coherent, justified, revised, or refined point among our judgments, intuitions, or principles by examining them against each other. In case we cannot reach a coherent point as such, that means that the disagreement about intuitions is just unreal or apparent, possibly due to linguistic reasons (Sosa 2009), deceptive hermeneutics (Sosa 2010), or verbal and normative (Sosa 2007c) issues.

Opponents of the epistemic status of intuitions might reply to this line of thinking that if the above-mentioned competence makes epistemic intuitions reliable, then in virtue of what does it do so? In other words, it might be said that Sosa stated that the aforesaid competence provides our intuitions with reliable evidentiality, but he did not tell us what it is which enables it to do so. This is not explained by the competence theory in terms of which intuition is supposed to be defined; they may object. However, a proponent of the competence theory may reply to this worry in two ways.

The first is to deny the objection: competence provides our intuitions with reliable evidentiality by virtue of attentiveness or the like. To this, it can be replied that attentiveness and the like are not part of the structure of epistemic excellence because false intuitions are attributed to performance errors in contrast to deliverances of competence, not to the attentiveness and the like. And since they are correctable by paying more care, attention, etc., through reflective equilibrium, this means that attentiveness and the like are external to the structure of epistemic excellence.

Sosa is required to explain that the structure of epistemic excellence supervenes entirely on factors that provide our intuition with its evaluative properties. Otherwise, Sosa's thesis will be understood as follows: epistemic excellence provides our intuitions with their evaluative properties, but it possesses none. In addition, he should show that introspective and sensory experiences do not supervene on factors that provide our introspections and perceptions with their epistemically evaluative properties. That is why, in introspection and perception, false intuitions are attributed both to performance errors as well as the system itself.

The second is to lend credence to epistemologists and their intuitions: epistemologists are competent, and their intuitions seem to be true. But this view has problems. Let us suppose, for the sake of argument, the existence of such epistemic excellence. It does not follow that it is the only generator of intuition even when attentiveness and the like are there. For example, one can have practical reasons that generate an intuition that does seem true. Suppose there is a competent epistemologist with attentiveness and the like, but he is a poor man. Suppose further that he has a Platonic intuition that knowledge is justified true belief. Imagine, for some reason or the other; a milliner offers him a million dollars for promoting an anti-Platonic epistemology. In spite of his attentive competence, a million dollars may generate an attraction to assent that knowledge is not justified true belief. In this case, what makes it true is not the subject's competency. Nor is it her attentiveness and the like. It is the million dollars which attract him to assent it.

The problem with this line of thought is that it practicalizes epistemology, reducing epistemic phenomena to mere practical factors. Whether belief should be considered knowledge is independent of practical factors such as interests, concerns, and so on. What is at stake may determine one's preferences. But epistemic states are defined by their role in explaining the character of warranted knowledge. And while the value of practical interest is appraised in terms of practical success or failure, the quality of epistemic phenomena is assessed in terms of the condition of being fit to contribute to attaining knowledge.

Still, we are without a good reason for supposing that Sosa's proposal is unproblematic. In general, the objection to Sosa's strategy can be raised in a different way as follows: Sosa's theory of epistemic competence does not specify a standard on the basis of which we can verify our intuitions. That is to say, Sosa's proposal is that the subject's epistemic competence is the source of both true and false intuition, but this account did not tell us whether competence as such helps in verifying our intuitions. Nor does it tell us in virtue of what it does do so when it does. In response, Sosa again might point out the virtues of reflective equilibrium. But even so, there are several problems with this view.

First, even if a reflective equilibrium is free from any objection, it is not part of the epistemic competence, as I have already argued. Second, suppose that reflective equilibrium is a reliable standard for verifying our intuitions; this does not mean that Sosa's epistemic competence is helpful in calibrating our intuitions. All it means is that the truth of intuition can be derived from a dialectical criterion. And, in such a case, it can be asked: is it that dialectical criteria need to be applied instead of epistemic criteria? The main difference between the dialectical and epistemic criteria lies in the conception of the argument contained in each.

In the dialectical criteria, the argument is acceptance-based. That is, the focus is on whether or not the premises and conclusion are subject to retraction in the context of a practical process of opinion revision. Suppose I debate the subject of knowledge with other people. I say it is a justified true belief, and they say it is not. The process of evaluating our opinions has two options: either it goes ad infinitum, or it ends up agreeing with one particular view. If this is true, the aim of the dialectical criteria is just to bring other interlocutor(s) into an agreement with one's point of view. Whether or not what was agreed on is true is not so essential. Indeed, the dialectical argument seems a familiar kind of argumentation that occurs in everyday conversations.

On the other hand, in the epistemic criteria, the argument is knowledge-based. That is to say; the epistemic argument focuses on whether or not the premises and conclusion are justified or known to be so. Suppose, for example, that we are trying to investigate the following proposition: Norman knows that the President is in New York City. In this case, we will be concerned with the following questions: is what Norman knows considered knowledge or belief? Is it a true belief? Is it a justified belief or just a matter of luck? Is having evidence enough for a belief to be justified? Understood more precisely, we will be concerned with questions that have to do with the creation of knowledge.

If so, we have a dialectical-epistemic dichotomy where the dialectical argument does not necessarily entail or imply the epistemic one. In addition, it has come to be more apparent that Sosa's competence theory is not an appropriate account of the epistemic value of intuition. However, Sosa's competence theory is not the only approach to the role of intuition in epistemology. Competing views are available.

3.2.3 Chudnoff on Phenomenal Dogmatism

Taking intuition to be the proper target of his analyses, the metaphilosopher Chudnoff addresses reliability and justification. Due to this fact, he discusses the epistemology of intuitions, i.e., how they can provide us with justification and reliable knowledge. In this context, he introduces what he calls 'Phenomenal Dogmatism,' according to which intuition, in virtue of its presentational phenomenology, prima facie justifies a belief. In his words:

> it is dogmatic because it includes the thesis: if it intuitively seems to you that p, then you thereby have some prima facie justification for believing that p. It is phenomenalist because it includes the thesis: intuitions justify us in believing their contents in virtue of their phenomenology—and in particular their presentational phenomenology (Chudnoff 2011b, p. 313).[10]

Chudnoff defends a phenomenal dogmatism about perceptual justification. That is, the "perceptual experiences prima facie justify us in believing propositions with respect to which they have presentational phenomenology and do so in virtue of their presentational phenomenology" (Chudnoff 2013a, p. 83). Just as with perception, intuition prima facie justifies us in believing propositions with respect to which they have presentational phenomenology and do so in virtue of their presentational phenomenology (Chudnoff 2013a, pp. 93–8).[11] Presentational phenomenology then is what explains the epistemic power of both perception and intuition as well.

What is more, he considers four skeptical challenges to this claim. First, the argument from inescapable error, according to which intuitions—specifically mathematical intuitions—are inescapably unreliable. To this, he replies that "error—specifically error at the level of the content of intuition—cannot only be avoided

[10] On Chudnoff's account, the phenomenology of intuitions is important both to their nature and epistemic status.

[11] It seems Chudnoff thinks that just because perception has presentational phenomenology, it does entail that intuition also has presentational phenomenology.

through further reflection, but also positively corrected" (Chudnoff 2013a, pp. 103–4). Second, the no access argument, according to which intuition is just an intellectual hallucination since it does not provide us with access to its abstract subject matter. Contrary to that, he replies that intuition provides access as such, i.e., awareness. In (Chudnoff 2013b), the main idea is that intuition amounts to knowledge in virtue of, at least partly, its ability to make the subject aware of its abstract object. Third, the argument from disagreement, according to which people undeniably have conflicting intuitions that have presentational phenomenology. To this, he responds that "disagreements that derive from conflicting intuitions might not derive from intuitions that have presentational phenomenology with respect to the conflicting propositions. And, we can add, even if some do so derive, they might not be persistent among normal people" (Chudnoff 2013a, p. 107). The fourth challenge is the argument from experimental philosophy, according to which "intuitions are unreliable because they are influenced by a variety of factors that, like cultural background, do not track the truth about their subject matter" (Chudnoff 2013a, p. 109). Chudnoff's brief reply to this challenge can be stated as follow:

> the experiments need to control for the conditions that must obtain in order for a judgment to derive from intuition ... Further, if the skeptical challenge is supposed to defeat the prima facie justification intuition experiences provide … then the experiments need to control for presentational phenomenology with respect to the propositions expressed in the answers to surveys (Chudnoff 2013a, p. 113).

Unless experimental philosophers fulfill these conditions, they will never be able "to account for the motivations philosophers give for their intuitions about cases" (Chudnoff 2017, p 384). Having replied to those challenges, he concludes that Phenomenal Dogmatism holds.

Next, he examines the main competitor to Phenomenal Dogmatism, namely, understanding-based views of intuitive justification, according to which "intuitions justify beliefs because they are based on understanding" (Chudnoff 2012b, p. 1).[12] That is, your intuition justifies you in believing a certain proposition in virtue of your psychological processes that depend on its content and way of the combination of its concepts to constitute your full understanding of it. Discussing two counterexamples, viz., The Jordan Curve Theorem, "a simple (i.e. non-self-intersecting) closed curve in a plane separates the plane into two disconnected regions—an inside and an outside" (Chudnoff 2013a, p. 125) and naïve set theory; Chudnoff argues that understanding plays a role in having intuition but not in establishing its justifying value. So, this view loses its attractiveness. Nevertheless, understanding might play a positive epistemic role in justifying intuition experiences. A role as such "depends on two conditions. First, there is the extent of the presentational phenomenology of the subject's intuition experience. Second, there is the extent of the information that is part of the subject's understanding" (Chudnoff 2013a, p. 144).

[12] Chudnoff, who challenges empiricism, also challenges the understanding-based account of intuition on which Bealer and Sosa have elaborated.

Chudnoff has been so far discussing one rational role of intuition, namely, intuition as a justifier. (Chudnoff 2014b) talks about one more role, which he calls intuition as evidence that is something that helps to prove that a belief is true. In addition to these two "methodological issues," as he calls them; (Chudnoff 2011c), he discusses a third rational role a phenomenal intuition plays, namely, its 'action-guiding role,' according to which

> in order to gain justification for believing the conclusion of an argument by inferring it from the premises in that argument one must "see" that the premises support the conclusion ... one's "seeing" such a support relation consists of one's having an intuition (Chudnoff 2013a, p. 146).

So, although intuition's guiding and justifying roles are different, it performs its guiding role by virtue of having the justifying one.

What is left is to discern in virtue of what intuitions enable us to get knowledge. His response to this question is that just as with the ground of perceptual knowledge, the ground of intuitive knowledge lies in that the veridical presentational phenomenology of intuition experience puts its subjects in a position to gain knowledge because it makes the relevant proposition intellectually seem to be the case and it makes its subject as if she is aware of a truth-maker for the relevant proposition. In sum, like perception, intuition grounds knowledge without necessitating it (Chudnoff 2013a, pp. 174–207). What about intuitive awareness of abstract objects?

On his view, perceptual experience makes its subject aware of a truth-maker for the relevant proposition because: [1] her awareness is dependent on the object of experience, and [2] the presentation of the object differentiates it from its sensory background. In a slightly different way, intuition experience makes its subject aware of a truth-maker for the relevant proposition because: [1] her awareness is constituted as non-causally dependent on its abstract object, which along with other intuition experiences constituents, such as thoughts and imaginings; instantiates and thereby determines the phenomenal character of intuition experience, and [2] constituents as such are arranged in a way that the phenomenology of this experience differentiates its abstract object from its background (Chudnoff 2013c). And this is the significance of presentational phenomenology.

These are the central arguments presented by Chudnoff to defend the epistemic significance of intuition. Let us consider Chudnoff's reply to skepticism about intuitions. Skepticism about intuitions is the view that intuitions are unreliable as a source of knowledge simply because they are unstable and are influenced by irrelevant factors to their truth. Chudnoff's response is that skeptics, while talking about intuitions, do not take into consideration the fact that intuitions have presentational phenomenology.

But this characterization raises a number of tricky questions. First, how do we determine whether an intuition made in an epistemological text is accompanied by such a presentational phenomenology in the relevant group of subjects? Those who claim that the epistemic status of intuitions can be explained by their alleged presentational phenomenology should be deeply concerned with how to detect the presence of an intuition accompanied by such a presentational phenomenology. Second, if it is accompanied by such a presentational phenomenology in the relevant group, how do we determine whether its presence is effective and it is not just an idle accompaniment? Third, if its presence is effective and it is not just an idle accompaniment, Chudnoff did not show why intuitions with presentational phenomenology do not conflict and vary according to different factors. Chudnoff might say that he is talking about intuitions with presentational phenomenology being veridical. To my mind, this response is worth exploring further. That is, it, in turn, leads to another question about how to assert whether or not an intuition with a presentational phenomenology is veridical. We need to decide the valid and authoritative standards under which an intuition with a presentational phenomenology is veridical. To date, such standards are not provided and accordingly, Chudnoff's proposal of veridicality is just an assumption of the truth of intuition. In light of the above-mentioned queries, while I can see the motivation for this kind of suggestion, I think it raises more difficulties than it solves.

Setting these worries aside, the question arises as to how Chudnoff explains the phenomenology of intuitions in virtue of which they are epistemically significant. Chudnoff's basic strategy places much weight on an assumed analogy between intuition and perception. Both have the putatively characteristic phenomenology. I have already argued that this supposed analogy instantly raises many worries that are not easily accountable for. Yet, some philosophers, Chudnoff included, think that this analogy is legitimate. Even reliablists, such as Ernest Sosa, repeatedly tried to save the reliability of intuitions on the basis of drawing a similar analogy. That is, if the error is enough reason not to consider something a reliable source of knowledge, then this standard should be applied in the case of perception and intuition as well. For we all know that sometimes our senses might wholly deceive us, and so intuitions do. If this is so, then either we are justified in relying on both of the faculties, or they both are not reliable.

But this is not the whole story. This line of argument focuses only on the possibility of an error on both sides: intuition and perception. But it ignores the most important thing, which is the possibility of checking out the error. That is, with perception, it is easy to identify and inspect the shortcomings. With intuition, this is not the case. Accordingly, one might object that "worries about intuitions are more pressing than those about perception" (Talbot 2009, p. 158). For with perception, we know which circumstances render perception reliable or unreliable. For instance, we know that the un/reliability of a perceptual judgment depends on whether the light is on/off. This aspect is not available in the case of intuition. Put simply, "we don't know what is the parallel for intuition of making sure that the light is on" (Swain et al. 2008, p. 148). This is where the aforementioned analogy fails, and therefore intuition and perception are disanalogous. Chudnoff cannot easily close his eyes to these points. It is open to him to keep defending his view about the phenomenology of intuitions. But unless he meets this challenge, defending his view on the basis of the aforesaid analogy becomes difficult.

So far, I have been arguing that the claim about the phenomenology of intuitions is imprecise, and the intuition-perception analogy does not hold. Yet, let us suppose for the sake of argument that Chudnoff is right. Intuition is analogous to perception.

And both have a characteristic phenomenology. Still, the question may arise: why is it necessary that an intuition having phenomenology entails its being true? One might worry that it is an exaggeration to say that one's intuition "cannot be true unless that phenomenology is present" (Cappelen 2012, p.80). What is true can, or perhaps must, be true regardless of whether or not it has phenomenal character. Suppose that someone intuits that 1 + 1 = 2 and that this intuition is true. This does not mean that it is true in virtue of its phenomenal character. On top of that, having a particular phenomenology is simply a property. If this is so, Chudnoff does not focus on things themselves but on their properties. And it is inappropriate to say that the truthfulness of something hangs on having a certain property. For instance, the truthfulness of 'x is a teacher' is independent of x's being a great researcher. The property of knowing Arabic, for example, has nothing to do with the truthfulness of 'Manhal is a smart PhD student.'

This is not to say that intuition is not true because the claim that it comes with a certain phenomenology is incorrect or because the assumption of the intuition-perception analogy fails. On the contrary, the claim that intuition is true may be correct even if intuition has no phenomenal character, it might also be correct even if the view of the intuition-perception analogy fails. In other words, it is not necessary for the truthfulness of intuition to be established on whether or not intuition has a phenomenal character and the view of intuition-perception analogy holds. It might very much be established on something else or even on its own. If so, the question again arises what that thing in virtue of which intuition is reliable evidence? In what follows, we will see how experimentalists neglect any positive response to this question while Steven Hales tries to respond to it positively.

3.3 The Challenges of Experimentalism

Reliablists' position on the epistemic status of intuitions has received much criticism from experimental philosophy. Some philosophers like Weinberg, Machery, Nichols, Stich and many others have started raising doubts about the reliablists' thesis that intuitions constitute evidence for or against epistemic thought experiments.[13] In particular, I talk about experimental critics, who challenge the evidentiary

[13] It is worth mentioning that (Bengson 2013) responds to these recent empirical studies. For him, they do not undermine the epistemic status rationalists used to grand to intuitions simply because they rest on "the Answers-Express-Intuitions Thesis" (AEI) (p. 504). According to this thesis, "subjects' prompted answers express subjects' intuitions" (p. 503). Bengson argues that AEI is false. To that end, he examines four empirical studies that have been used "as evidence that intuitions about knowledge are culturally relative … and that epistemic … intuitions are vulnerable to ordering and wording effects" (p. 496). Bengson provides two observations about the subjects' prompted answers in such studies: "the first observation records the possibility of blind answers (answers in the absence of intuition)," and the "second observation records the possibility of stray answers (answers that depart from intuition). Answers do not always express intuitions, and intuitions are not always expressed in answers. In both directions, we find a gap between intuitions and

status of intuitions by a variety of arguments.[14] In what follows, I will be exposing and commenting on two main types of arguments provided by experimental critics of the evidentiary role of intuition: the phenomenon of variation in intuition and the unamenability to calibration of intuition.

3.3.1 The Variation Argument

According to experimentalists, reliablists' epistemological views regarding the epistemic status of intuitions are ill-grounded. However, the structure of the variation argument is as follows. If intuition varies, then it does not qualify as evidence. The findings of empirical studies support the claim that intuition does, in fact, vary both between and within subjects. These results of recent studies might have their roots in William James' pragmatism.[15] (James 2000)'s picture was one in which philosophers' "temperament" has a tremendous and inescapable effect on their philosophical approaches. James holds that philosophers have varying temperaments: some are "tender-minded", and others are "tough-minded." Those who are tough-minded philosophers have a certain characteristic of tending towards empiricism, and those who are tender-minded philosophers are favorably disposed towards rationalism.

As a result, experimentalists concluded that intuition does not serve as evidence. To show how experimental philosophers offer support for variation in intuition, I present the results of some empirical studies of intuition. A study by (Weinberg et al. 2001) finds that Western subjects do not consider Gettier's and Truetemp's scenarios as cases of knowledge, while East Asians do. It is concluded that intuition varies according to the cultural background of the subjects. The same results were emphasized by (Machery et al. 2004), in which Westerners' and East Asian subjects' intuitions about Kripke's Gödel-Schmidt thought experiment differed from each other. As a consequence, folk intuitions about reference do also vary from

answers. Consequently, [AEI] is false: subjects' prompted answers need not express subjects' intuitions" (p. 511). For example, in the order effect studies, "subjects' prompted answers ... expressed, not intuitions [about knowledge], but rather (order-influenced)" and "(background-influenced)" in the cultural relativity studies. That is, in these two cases, subjects' prompted answers expressed "hypotheses, guesses, or inferences." In the wording effect studies, subjects "expressed (description-influenced) emotional reactions," and "conclusions—inferences" in the affective bias studies, rather than intuitions (pp. 516–521). If so, if the empirical studies invoke that subjects' prompted answers express, not intuitions, but other mental states "that rationalists and their opponents can mutually agree have no positive epistemic status," then AEI does not give rise to any negative conclusions about the epistemic status of intuitions. On the contrary, it may "provide substantial prima facie motivation for the conclusion that the relevant empirical studies may actually support rather than undermine intuition's epistemic status" (pp. 521–3).

[14] Experimentalists, who question the traditional methods of analytic philosophy, also challenge the evidentiary status analytic philosophers used to grant to intuitions.

[15] Thanks to an anonymous referee for raising this concern.

culture to culture. The preceding two surveys argue that we are left with no option but to accept and support a naïve form of relativism, namely, ethnocentrism, i.e., to say that a particular intuition is false because it has emerged in a certain group of people or to say that a certain ethnicity has better intuitions than others. But what is true is supposed to be so regardless of the ethnicity in which it arises. In view of that, they conclude that intuition should not be considered evidence for epistemological claims.

There are other reasons that give an additional boost to the theory. Experimentalists engage in more studies to confirm that culture is not the only factor according to which intuition varies. To favor that view, (Swain et al. 2008) conducted a study by trading on purely empirical data and found that intuitions also vary according to the order in which cases are presented. For example, they noticed that participants' intuition about whether or not the subject in Truetemp's case knows the right temperature varies according to whether or not they were presented with either a clear case of knowledge or non-knowledge before presenting the Truetemp case. Furthermore, the empirical data shows other factors behind variation in intuitions, such as educational background (Kuntz and Kuntz 2011; Weinberg et al. 2001), native language (Vaesen et al. 2013), gender (Buckwalter and Stich 2013; Adleberg et al. 2015; Seyedsayamdost 2015), individual differences (Nichols and Ulatowski 2007), age (Colaço et al. 2014), moral judgments (Knobe 2010). This list goes on and on. All in all, if the above examples are true, it follows that the most recent kind of skepticism about intuition came to light within the movement of experimental philosophers who hold that disagreement in intuitions is real and, accordingly, intuition cannot be regarded as a reliable source of evidence in epistemology.

3.3.2 The Calibration Argument

According to experimentalists, the above widespread disagreement in intuition also suggests that intuition cannot both have independent calibration and remain epistemically useful. The second argument then holds that intuition seems to be offering resistance to calibration. This argument was first inaugurated by (Cummins 1998).[16] Cummins argues that scientists consider observation as a key source of evidence for their theories. A proponent of the reliability of intuition will consider the role intuition plays in epistemology to be similar to the role observation plays in science. That is, intuition is epistemologists' key source of evidence for their claims in thought experiments.

Contrary to that, the role intuition plays in epistemology is not analogous to the role observation plays in science. Observation, in science, is possible to be validated. Before accepting the telescope as a source of evidence, it was subjected to calibration. But, in most cases, it is impossible to calibrate intuition. That is because

[16] Cummins is then a naturalist or empiricist in Bealer's sense of naturalism or empiricism.

there are no independent means by which we can measure the accuracy of intuition. Therefore, the intuition-observation analogy is misleading and, accordingly, intuition, in most cases, cannot be deemed as a reliable source of evidence.

The calibration argument has been defended by (Weinberg 2007) and (Weinberg et al. 2012). In brief, the structure of the calibration argument is as follows. Something can be deemed as a reliable source of evidence if it can be calibrated. Intuition cannot be calibrated. Therefore, intuition is not a reliable source of evidence. Weinberg argues that sources of evidence, such as observation, perception, etc., are fallible, but validating them is always available. If a microscope gives two different readings at different times, we can always identify what affects it. After that, we either fix or restrict it. If none of these options is available, we doubt its reliability. But, fallibility in the case of intuition is a hopeless case simply because intuition cannot be calibrated against independent data and, accordingly, it cannot be considered a reliable source of evidence.

Of course, this is the area of experimental philosophy whose epistemic findings may have much to contribute to the ongoing debate about the epistemic status of intuition. Yet, one can always reply to experimentalists' claims by arguing that appealing to such types of surveys in order to show variations in people's intuitions can always be challenged by referring to exceptions or different results generated from different surveys. That is, one may refer to exceptions from the empiricists' surveys to conclude the possibility of different results. For instance, following the same data collection method as the experimentalists' studies, (Wright 2010) conducted a survey on paradigmatic cases, which shows "that only some intuitions (that is, intuitions about certain sorts of cases) are vulnerable to intuitional instability and that people are implicitly aware of which cases these will be" (p. 492). Or one may conduct researches that indicate the agreement among people's intuitions about some question or the other regardless of factors such as gender, culture, age, and so on. For example, a study of (Kim and Yuan 2015) is a study of "data [that] provide strong evidence that the purported cross-cultural difference in intuitions about Gettier-style cases does not exist" (p. 355).

That being the case, the experimentalists' view itself cannot be substantiated by empirical evidence because it cannot provide empirical evidence to support or prove the justification of its own allegations through its own method. If people's intuitions typically vary according to the above-mentioned factors, it is not easy to see how experimentalists' thesis about intuitions can be empirically confirmed. Furthermore, one might also criticize experimentalists for trying to make a transition from a few studies to a huge statement. In this context, one can understand Matthew Liao's observation that it is odd for experimentalism "to be making such a large generalization from very limited study" (Liao 2008, P. 4).

What is more, some might argue that experimentalists' findings commit are self-defeating. Experimentalists assert that if something is true, then it must be universally shared. (Kim and Yuan 2015; Wright 2010) show that experimentalists' findings are not universally shared. Therefore, experimentalists' criteria are self-defeating.

This endangers the empirical account in two ways. One is that experimentalists do their surveys in order to prove the unreliability of intuitions based on their instability. This presupposes that they completely trust the authenticity of the intuitions presented by the subjects who participated in their surveys. "Hence, they believe the spontaneous judgments they have collected do qualify as intuitions" or authentic intuitions (Liao 2008, P. 5). Second, if experimentalists' method is efficient and their results are empirically defensible, then this means that the same types of experiments may amount in future to completely different results since people's intuitions are unstable. For example, if people's intuitions are unstable, then the result that "there are large and systematic differences between East Asians and Westerners" may collapse in the future (Weinberg et al. 2001, P. 436). Thus, if this empirical method itself is efficient, then its results will be unstable, and the instability of something is enough reason not to believe in it. Therefore, it is self-defeating and, accordingly, false. And if this method is not efficient, then, due to this inefficiency, it is certainly better that people should not believe in it. So, either way, it is better that people do not believe in it.

What is more, the conflict of intuitions alone is not a sufficient reason to deny any epistemological role to them. In Sosa's words, "disagreement need not be substantive" (Sosa 2007c, P. 100). Moreover, a proponent of the evidential status of intuition may reply to the variation argument by saying that insofar, in epistemology, what is reliable is so regardless of what majority of people may say, the reliability of intuition in question does not depend on whether or not it is widely shared. If people agreed that $1 + 1 = 4$, it does not mean that $1 + 1 = 2$ is wrong. Indeed, people may agree on unreliable things and regard them as reliable. For example, humanity believed for a long that the sun rotates around the earth. If so, universality is not necessary to explain the reliability of intuition.

More to the point, experimentalists pose their argument as follows: something can have a positive epistemic status if it is stable and universally shared. Intuitions are neither stable nor universally shared; then they are unreliable, and therefore they have no epistemic role. However, the entire history of philosophy in general and epistemology, in particular, was established based on similar types of conflicts, such as the conflict between empiricism and rationalism, but none of them has found in the conflict itself a sufficient reason either to deny other's epistemic role or to assert its own. On the contrary, the conflict as such has formed the history of epistemology.

Further, the conflict of intuitions exists between empiricists and rationalists themselves. Hence, the conflict of intuitions and their alleged instability alone is not enough to deny their role in epistemology simply because being unstable and being epistemologically useless are two different things. In addition to that, one can say that conflicts, variations, contradictions, and so on form the matrix from which epistemology started and continued its progress and therefore, an appeal to the absence of conflict among intuitions can be considered as demand for stopping this progress.

However, suppose that experimentalists are right: for intuition to be reliable, it needs to be universally shared. Suppose an experimentalist forms the following hypothesis: intuitions vary, and then they are unreliable. With the empirical foundations in mind, a hypothesis as such is to be taken innocent until proven guilty. The experimentalist forms a thought experiment in which participants are asked whether the subject of the thought experiment knows or just believes. Suppose the results are as follows: 30% of Indians said she knows, and 60% of Americans said she just believes. The experimentalist concludes that intuitions vary, and then they are unreliable. Nevertheless, the result shows a different assertion, too: 70% of Indians and 40% of Americans do share intuition about the relevant issue. Such an intuition must be reliable according to the experimentalists' standard itself. This is what was missed from the experimentalists' view and what should be accepted by them.

On the other hand, experimental philosophers are required not to exaggerate the epistemic significance of their findings. According to these findings, intuition does not calibrate. Here, I will make two quick points. First, we are by no means convinced that the data collection method –experimentalists' method- will be an admitted method to judge whether or not intuition qualifies as a reliable source of evidence in epistemology. If, in science, observation does not calibrate, then a telescope can no more be considered an optical instrument to collect images of distant objects. The same story goes for intuition.

Second, if intuition does not calibrate, this does not entail that intuition is not a reliable source of evidence. What this means is that we do not know whether or not intuition qualifies as evidence simply because we cannot know what intuition is reliable and what is not. If observation does not calibrate, this does not mean that it is not correct evidence in science simply because we lack the criteria/telescope which plays the role of an optical instrument. What this means is that we do not know whether or not intuition is reliable evidence.

As can be noted from the above discussion, there is a marked diversity of views on questions of the epistemic status of intuition. Some philosophers move from disagreement to perspectival relativism. The central claim of this view is that between reliabilism and experimentalism, there is a middle ground which should not, I think, be ignored.

3.4 Perspectival Relativism About Intuition

Another position on the epistemic status of intuition is perspectival relativism, according to which the truth of intuition is perspectively dependent on the epistemic method in virtue of which one gains one's belief about the relevant intuition. However, this view differs from both reliablists' and experimentalists' approaches. It differs from reliabilism because it defends the claim that a view is only *relatively* true, depending on the use of basic, non-inferential and rational intuition. This means it agrees that rational intuition is a source of evidence, and this is enough for it to be different from experimentalism. Yet, it agrees with experimentalists since it defends the view that the relative truths depending on the use of basic, non-inferential and rational intuition should be brought into reflective equilibrium.[17] Defenses of perspectival relativism then are motivated by an attempt to find a middle ground between reliabilism, the view that intuition epistemically qualifies as reliable evidence of knowledge; and experimentalism, the view that intuition is epistemically useless as evidence of knowledge.

This type of view was championed by (Hales 2006). Here is the structure of his argument. There are three methods for obtaining epistemic beliefs about propositions. They are intuition, Christian revelation and the ritualistic usage of hallucinogens. No method is epistemically superior to any of the other two alternatives. Accordingly, although the results of these methods will be inconsistent, their epistemic status will be the same; because they will be relative to the perspectives or methods by which they were obtained. Therefore, amongst nihilism, according to which we have to neglect the existence of proposition, skepticism according to which knowledge is impossible, and relativism, perspectival relativism is true.

As mentioned earlier, Hales considers three methods through which epistemology can address its concerns. The first is intuition which, like Bealer, he differentiates between two types: physical intuition, which creates "beliefs about contingent truths," and rational intuition, which "generates beliefs about putative necessary truths" (Hales 2006, p. 20). Furthermore, while the reliability of physical intuition can be checked by appealing to empirical data, the reliability of rational intuition is examinable "through appeals to possible cases and nonempirical counterexamples" (Hales 2006, p. 24). But how is rational intuition reliable? To this question, he responds by adopting a view which holds that the "reliability of intuition is a task that intuition itself must perform" (Hales 2006, p. 27). It is easy to see that this leads us to the problem of epistemic circularity: the reliability of intuition as a source of justification relies on the very source (i.e. intuition itself).

The simplest example of this circularity is the case where he proposes a modest form of foundationalism according to which "there are justified propositions whose justification depends on nothing other than themselves" (Hales 2006, p. 32). So, he argues that rational intuition provides this illuminating sort of justification. Additionally, rational intuition produces basic, non-inferential and a priori beliefs. Unlike the Cartesian tradition, rational intuition is not indubitably certain, even if it produces basic, non-inferential, and a priori beliefs. Rather, it is defeasible because it is testable against a wide reflective equilibrium with different theories from different fields (Hales 2006, pp. 36–43).

The second and third methods "are Christian revelation and the ritual use of hallucinogens" (Hales 2006, p. 50). Those two methods produce epistemically foundational, basic, non-inferential and a priori beliefs that are also testable against reflective equilibrium. Hence, all three methods supposedly possess equal dependability. But the problem is that they yield inconsistent results. In a situation as such, we face the problem of judging which method is preferable. According to him, to

[17] This means that it partly differs from both reliabilism and experimentalism, though it partly shares a common ground with both of them.

solve this problem, we have to believe in one of the following views: skepticism, nihilism, or relativism. Skepticism is the view that philosophical knowledge is impossible. He rejects skepticism simply because it "is not merely self-destructive; it leads to the knower paradox" (Hales 2006, p. 94).[18] Nihilism is the view that philosophical propositions do not exist. He rejects nihilism simply because it proposes that we must "reject philosophical propositions as either unphilosophical or not really propositions (Hales 2006, p. 95). We are left with only one option, relativism; the view that "what propositions are true is therefore dependent on, and relative to, method" (Hales 2006, p. 96).

He considers the well-known anti-relativism argument, which is sometimes called a self-refuting argument. To meet this challenge, he distinguishes between global and perspectival relativism. He accepts that "global relativism," according to which "everything is relative," is self-refuting because the modal proposition "everything is relative" can be true either absolutely or relatively. If it is absolutely true, then it is true in all perspectives, and accordingly, relativism is false. If it is relatively true, then absolutism must be true, at least from some perspective, and so relativism is false. Either way, relativism is false. In view of that, he proposes perspectival relativism that takes the form that "everything true is relatively true" (Hales 2006, p. 102). Here, the modal proposition "everything true is relatively true" absolutely won't be a problem for any proponent of perspectival relativism since there is no non-perspectival truth. Therefore, the self-refuting argument is sidestepped.

He goes on to defend his version of relativism against naturalism that reduces propositions to science. To that end, he draws a distinction between two kinds of naturalism. The first is "global naturalism," according to which knowledge is possible only by using the scientific (natural and social) methods of solving philosophical matters. In reply, he says that it is not true that non-naturalistic methods provide no knowledge at all. In view of that, global naturalism is wrong.

The second is "local naturalism", the view that knowledge acquired by non-naturalistic methods such as intuition is untrustworthy (Hales 2006, p. 167). He also distinguishes between two versions of local naturalism. One is Goldman's continualism, according to which rational intuitions, "what seemed like a priori philosophical propositions are empirical and a posteriori after all … Thus rationalism is both continuous with science and superior to revelation and hallucinogen" (Hales 2006, pp. 167–68).

In reply, Hales argues that except for British empiricists, who believe in the superiority of experience over reason as a source of knowledge, philosophers often used to depend on pure reason without trying to emphasize disagreement with science. In addition, "both Christianity and hallucinogenic traditions have been influenced by science" (Hales 2006, p. 168). Another form of local naturalism is the view of experimental philosophers who used the empirical method of social science to conduct studies which show that rational intuition is unstable because it varies

[18] Knower paradox belongs to the family of self-referential paradoxes.

according to factors that has nothing to do with the truth of intuition itself, such as culture, language, socioeconomic background, etc., and accordingly intuition is not a reliable source of knowledge. To this, Hales responds that rational intuition is a subject to reason in the wide process of reflective equilibrium. Naturalists should explain why mathematics, pure logic, etc., do not have to be empirically testable, but rational intuition does. They are also required to show that philosophers' intuition is not better than non-philosophers'. What is more, if intuition does not imply its truth, experimental philosophers should justify their dependency on the standard of social science by which they judge or decide the value of intuition. Otherwise, they may be accused of presupposing that normativity is true, but cognitivism is false. If a justification as such is not offered, then knowledge can be obtained by some other methods. Rational intuition may be one of these methods without being superior to them (Hales 2006, pp. 169–77). Therefore, experimental philosophers are wrong.

Having explained and presented Hales' approach, let us consider some of its problems. Let us start with his initial thesis that rational intuition is a method on which philosophy relies while addressing its issues. This claim seems to be objectionable. It might be said that it is argument and not rational intuition that philosophy is all about. That is, what a philosopher does is to argue (i.e., to give a reason or set of reasons with the aim of persuading others) that a view, theory, or idea is true or false. For that reason, if the aim of philosophy is to remove uncertainty and ambiguity so that we are left with clarity in terms of what is right and what is wrong, then philosophy is basically motivated by different forms of argument.

Next, in order to establish his perspectival relativism, Hales suggests that there are three methods that produce beliefs about philosophical propositions: rational intuition, Christian revelation and ritual use of hallucinogens. Here, he is required to do two things. He first needs to argue well that knowledge is beliefs about philosophical propositions. Further, he has to show why Gettier is wrong in arguing that justified true belief is not knowledge. And he has to show that Christian revelation and ritual use of hallucinogens have something to do with knowledge about philosophical propositions. Just as he criticizes naturalists for their dependency on scientific methods dealing with unempirical propositions, he himself can be criticized for his dependency on religious methods dealing with unreligious propositions. Since methodic divergence is the bedrock of his form of relativism, he needs to consider these two essential comments.

However, these three methods yield conflicting results. Just as none of the methods is at a higher level than another, so too the result of any of these methods does not have any truth that is superior to those of another since all of them will be submitted to the authority of reflective equilibrium which is, according to Sankey, "a method of justification whereby judgments about particular cases are brought into conformity with general principles … [and] in which a broader range of considerations is taken to be relevant" (Sankey 2014, p. 288).

If we apply this quotation to what Hales is saying, at least regarding intuition, we would say that an intuition that was already subjected to many considerations in the process of reflective equilibrium is true. As if reflective equilibrium necessarily leads to truth. But that is not necessarily the case. It might be objected that reflective equilibrium is very weak to lead to truth (Kelly and McGrath 2010). Instead, they argue that it might lead to what (DePaul 1986) called epistemic foundationalism. Suppose that Hales is right. Rational intuition generates foundational or self-justifying beliefs. Then, if two different people have two different intuitions that produce two different initial beliefs, both beliefs are supposed to be foundational or self-justifying. The danger is that there is no guarantee each of these two people will not fully trust their initial belief even after being subjected to many considerations in the process of reflective equilibrium. That is, these two beliefs might not be brought into conformity with general principles because these two people may not reach equilibrium because both will keep full trust in their foundational, self-justifying, and initial beliefs. In such a case, maybe all that reflective equilibrium can do is confirm their initial beliefs. And this leads to what (Brandt 1979) calls conservatism, in opposition to the quality of not trusting change. In view of that, it seems as if Hales just took for granted that reflective equilibrium is to be relied on as a source of truth. If reflective equilibrium is unreliable, then the base upon which he rests to defend relativism is no more available. Therefore, relativism is wrong.

Last but not least, two comments are left. One is that, unlike Hales, I do not think that Goldman argued against the a priori status of intuitions. Nor does he defend their a posteriori status. What Goldman says is that intuitions as evidence "are a clear example of a priori cognition" or "non-experiential and ratiocinative cognition." This depends on "what cognitive science tells us about the cognitive processes underlying classification intuitions." But whether there is evidence that they have this evidential status is an empirical question (Goldman 2010, pp. 117–41). Therefore, Goldman was trying to distinguish between what is a priori and what is a posteriori in intuitions. Put simply, the statement 'intuition is evidence' is a priori. But to wonder whether there is evidence that 'intuition is evidence' is an empirical matter, and we have to wait for what cognitive science may tell us regarding it. I think this is different from saying that Goldman argued for the a posteriori status of intuitions.

Another is his adoption of modest foundationalism about intuition as a way of knowing (Hales 2000, 2006, 2012). But when he defends his perspectival relativism, he utilizes the thought of many philosophers such as Rorty, Nietzsche, Kuhn, etc., who are traditionally known not only as relativists but completely anti-foundationalists. For example, to the best of my knowledge of his works, foundationalism for Rorty is the philosophers' constant fault from Plato to his time, but it is no longer acceptable. This is the main idea in (Rorty 1979).

I have shown so far that there are several ways to challenge the preceding three accounts of intuition's epistemic status. Yet, while I may agree that the epistemic status of intuition is often in question, it is not something we can so easily dismiss. What follows is an attempt to show how.

3.5 The Desiderata for an Account of the Epistemic Status of Intuition

Having explored and undermined the main positions on the epistemic status of intuitions, now it is time to articulate what seems to me to be the right position that any adequate account must respect. The suggestion is that intuition alone has no epistemic status in the way considered by these positions. This is true in two senses. One, as I have shown in the second chapter, that intuition is not propositional, and hence it has no propositional content (i.e., it is not sensitive to the truths-conditions under which a given proposition is true or false). Second, being non-propositional makes it also impossible for it to be part of a logical argument. However, the only way in which intuition can have an epistemic status is to be evaluated within a different kind of argument which in turn should be appraised differently.

To be precise, intuition appears to be a crucial part of the relevant thought experiment. The thought experiment offers an argument. By 'argument', I understand its epistemic form. That is, it appears as something acting as a justification for some position. Therefore, the question of epistemic status does apply to the argument and not intuition. The question is how intuition which is not propositional, can be a crucial part of an argument and what is a good argument. That depends on what is meant by 'argument'. I shall be arguing that epistemic argument is structurally quite different from the logical argument, and that is why intuition can be part of it though it is non-propositional. If so, if both of the arguments are different, then both the criteria and appraisal words are also supposed to be different. I propose that the evaluation of arguments associated with the field of epistemology is best understood in epistemological terms. Here are the desiderata.

3.5.1 Intuition Appears to Be a Crucial Part of the Relevant Thought Experiment

As indicated in the first chapter, epistemic intuition is genuinely elicitable by thought experiments (Colaço et al. 2014; Dennett 2013; Machery 2011; Gendler 2007; Brendel 2004). For each of the examples discussed there, intuition captures something important about its fundamental structure. There may be several possible ways to interpret thought experiment in a way that fit this claim. Here is one: philosophers explicitly make use of intuitions to construct their thought experiments or arguments. Thus, the charitable strategy I follow in such a situation is a non-removal strategy.

Simple Non-removal Strategy In almost every case, one cannot remove the relevant intuition and have no significant effect on the essential constituent, lucidity and argumentative strictness of the argument. Therefore, intuition is a particular kind of matter of which a thought experiment or an argument consists and which has a tangible, solid presence.

By way of illustration only, and not by way of limitation, what follows are examples of non-removable intuition as a pivotal part that, combined with other pieces, makes up the whole, i.e., thought experiment or argument.

A typical case of that is Jackson's thought experiment on Fred. Just try to consider the Fred case without intuition: our having all the physical information about the subject won't help us to know everything about his color experience. That is to say that even after we know everything about his brain processes, physiology, history, body, etc., something will still be left out. In case of removing this intuition, the thought experiment will make no sense. Thus, the worse thing one can do is to remove the intuition involved in the thought experiment and keep what remains.

In the same way, the intuition that the subject's previous physical knowledge is incomplete serves an argumentative function in the text of the Mary case. It serves to complete the process of transitive reasoning that grounds the conclusion that 'physicalism is wrong.' That is why, after releasing her from her black-and-white room, she will acquire more than all the physical information already known to her.

The same strategy works well in Searle's Chinese Room case. It became clear, given the intuition that the computer (the protagonist of the story) lacks cognitive states, that the claim of artificial intelligence that an adequately programmed computer can possess cognitive states just like the human mind is wrong. In a like manner, in Chalmers' Zombie case, the thought experiment could make sense by virtue of the intuition that the realm of Zombie or Zombie itself (the subject of the story) is, indeed, conceivable. The same is true, I suspect, in a very wide range of cases. In Cohen's Lottery cases, it is clear, given the intuition in case 1 that the subject does not know that she will lose and in cases 2 & 3 that the subject knows that she will lose, that induction is not an unfailing ground for knowledge.

Paradigmatically, Lehrer cannot arrive at the conclusion that external reliabilism is wrong without going through the intuition that the subject Mr. Truetemp, does not know what the temperature is when he truly says what it is. Nor can Goldman conclude that our assessments change depending on the distinguishability or discriminability of the situation without the ground of his two intuitions that are 'Henry knows' to 'Henry does not know'. The same story goes for Gettier's Smith-Jones example in which his intuition that Smith does not know the proposition 'the one who will get the job has ten coins in his pocket' is true justifies his conclusion that the analysis of knowledge as justified true belief sounds false. With the minimal revision, one can see that without the intuition that we do not refer to Schmidt when we use the name Gödel, Kripke's Gödel–Schmidt thought experiment would lose much of its argumentative force. Last but not least, without the move from the intuition that the subject is epistemically unjustified in believing that the President is in New York City to the conclusion that external foundationalism is problematic, BonJour's example of Norman will lose all of its argumentative significance.

The Non-removal Strategy will work, I suspect, if not in all, then at least in a very wide range of cases. All we have done is a careful study of the uses of intuition which reveals that intuition often plays a significant argumentative role and that one cannot remove it without negatively affecting the overall argumentative transparency.

3.5.2 Thought Experiments Offer an Epistemic Argument

The authors of thought experiments just considered present them as ones where intuition seems to be part of a process of argument. Suppose that each of the examples above is structured as an argument; the question arises, why would this be? The answer depends on what an argument is and whether these examples are demarcated by this conception. The word 'argument' was defined by Monty Python's famous sketch "Argument Clinic" as follows. "An argument is a connected series of statements to establish a definite proposition." One of these propositions is called the conclusion, and the rest of them are called premises. In (Goldman 1994, p. 27)'s words, "an argument is a set of sentences or propositions, one designated as conclusion and the remainder as premises". Is any list of propositions an argument?

The answer to this question, as this definition indicates, is no. In addition to such a list, language is the material out of which an argument is made. But even if we use language to make a list of statements, this list won't be an argument without some terms, as in "therefore" in the following example. Statement one: All human beings have the inherent right to life. Statement two: Ram is a human being. Statement three: Therefore, Ram has the inherent right to life. The term "therefore" points out that the statement that comes after it is a conclusion. So, arguments include terms called conclusion markers such as therefore, hence, thus, then etc.

But the previous example may also take the following form: since Ram is a human being, and all human beings have the inherent right to life, Ram has the inherent right to life. In this case, it is the term "since" that points out that statements one and two lead to statement three or the conclusion. So, the argument also includes terms called premise makers, such as since, because, for, as etc. And both conclusion and premise makers together are called argument makers because they turn a list of statements into an argument.

However, I propose that this formalization of argument applies to logic, not epistemology. Goldman supports this idea and writes:

> There are several viable conceptions of a good argument. One such notion, the 'logical' notion ... While this is one legitimate conception of a good argument, it is not satisfactory for all purposes ... an epistemological sense of good argument is (also) extremely important (Goldman 1997, p. 159).

It is true that both logic and epistemology investigate adequate ways of reasoning. But while logic is interested in propositional reasoning, epistemology is more interested in how such reasoning leads to knowledge. Accordingly, it seems that for us to learn more about the concept of argument, it is better to distinguish between two types of it: logical and epistemological.

Logical Argument it is one that consists merely of a list of propositions. How do the premises and conclusion affect the success of the argument? According to (Strawson 1952), a logician is interested in how statements are connected to each other. Suppose that we have found out all statements that are true about the world. And we draw out all the inferences we can from such statements. The logician only tells us whether what we claimed about the inferences – their validity - was really true. Were the inferences really valid? Were they really consistent? If so, that is good enough. If not, we have to figure out what went wrong. The logician is interested in how statements are connected both to each other and to the conclusion, regardless of whether or not the premises offer reasons why one should believe the conclusion. For example, if P implies Q, it does not follow that P is a reason for Q. Similarly, if "Humans are mammals" implies that "Humans are not reptiles", it does not follow the former is a reason for one to believe the latter. The argument, in this sense, takes proposition to be the primary truth-bearers.

There are three basic types of argument: deductive, inductive, and conductive arguments.[19] Deductive arguments are such that the truth of their premises guarantees the truth of their conclusion. Deductive arguments are either/or things. A good deductive argument gives us conditional certainty. A bad one tells us nothing. On the other hand, inductive arguments are such that the truth of their premises makes the conclusion more or less probable. This means that inductive arguments can be either weak or strong. An example of a strong inductive argument: the sun has risen every day in the history of the universe. Therefore, the sun will rise tomorrow. Example of a weak inductive argument: every time I have seen Marianne. She has been wearing earrings. Therefore, the next time I see Marianne, she will be wearing earrings. Lastly, in a conductive argument, the premises individually offer support for the conclusion (Govier 2010). For instance, your words were hurtful to her. She is an unselfish and courageous worker. Therefore, you should not have spoken so harshly. A conductive argument can be either cogent, as in the previous example, or not cogent.

Logicians study arguments by studying valid arguments forms, arguments that are valid in virtue of their forms as opposed to their contents. With relatively few exceptions, this is true for most deductive arguments. Let us have a look at the following two examples of a deductive arguments. For instance: All men are mortal; Socrates is a man; therefore, Socrates is mortal. Second instance: All actions that produce the GHGN (The greatest happiness of the greatest number) are right; that action produced the GHGN; therefore, that action was right. The form of those two arguments is exactly the same: All As are B, S is A, and therefore, S is a B. Yet, the content of the two arguments is quite different. The first is an argument about Socrates and mortality. The second is about morality and GHGN. The two

[19] Some logicians talk about more kinds of argument, such as mixed argument (i.e., it is built from both deductive and inductive arguments), argument from analogy, and so on. But I am not going to discuss them here because doing so does not serve the main purpose of this book.

arguments are about completely different topics and content, but each of the arguments has the same form.

Epistemic Argument it shares important features with a logical argument, such as language. Here also, language is the material out of which the argument is made. Epistemic arguments also share the idea of premises, conclusion and the types of argument with logical argument. But the focus here is not on premises and conclusion as a list of propositions or statements.

Instead, its center of interest is on the argument, which, as noted in the first chapter, does not consist of a list of propositions but epistemic reason(s) or evidence(s) produced for the claim that this or that theory is justified or unjustified. Reason(s) or evidence(s) produced within the argument are known either as premises or a conclusion (i.e., the assertion that the premises are provided). Accordingly, the primary bearer of values here is not a proposition but the argument itself. A serious question arises here what is it for a premise or conclusion to have a reason? "A reason is a consideration in favor of acting or representing the world in a certain way" (Jenkin 2020, p. 259). In other words, the epistemic reason is a basis for the belief in facts about the world. Given the fact that intuition is a mental state, as I have already argued in the second chapter, what does it mean for a state of intuition to have an epistemic reason as such? The answer given by (Jenkin 2020) is that "if a state is of a type that can be based on epistemic reasons, that makes it apt for epistemic evaluation" (258). As we will see later, intuition is capable not only for having reason to accept it, but it itself can be epistemic reason on which another state, premise, etc., can be based.

In epistemic arguments, however, the arguer takes the argument itself to be sensitive to evaluation. In this sense, I look at the argument as a tool used for the purpose of providing a good reason(s), evidence(s), or justification(s) to believe the conclusion. For example, Statement one: All human beings have the inherent right to life. Statement two: Ram is a human being. Statement three: Therefore, Ram has the inherent right to life. You can see that statements one and two offer good epistemic reasons for believing the conclusion.

Following this line, one may see that argument usually takes one of the following forms: deductive, inductive, or conductive. Those three types of argument are common to both logic and epistemology. Nonetheless, both logic and epistemology interpret them differently. We have already seen how logic interprets them. Accordingly, in what follows, I will reconstruct the thought experiments discussed in the first chapter as an epistemologically deductive, inductive, or conductive argument. I predict that more detailed research of more cases will find more of the same. I will also show how what I identified there as intuition captures an important part of their fundamental structure, namely, epistemic reason in the way above-defined.

The first kind of argument epistemologists have given the most attention to is the deductive argument (Sinnott-Armstrong and Fogelin 2013). Epistemologists often frame their arguments in a deductive way where the argument usually flows from the general to the particular. A deductive argument derives the acceptability of a conclusion from the acceptability of premises. So, an example you will find in many textbooks is the following. Statement one: All human beings are mortal. Statement two: Socrates is a human being. Statement three: Therefore, Socrates is mortal. You can see that if statement one and statement two are both accepted, statement three has to be accepted. They offer epistemic reason(s) for believing the conclusion. That is a good example of a deductively good argument. A good deductive argument is structured such that if the premises are true, the conclusion has to be so. But in an argument like this, most of the real work is done supporting the premises. So even though epistemologists often frame their arguments in terms of deductive proofs, most of the real work of epistemology is in supporting the premises, and this makes epistemology an evidential discipline. We look for evidence: are the premises proper?

Suppose someone has committed a murder. The police officer comes in to investigate. To that end, the detective starts picking up small pieces of stuff which can help to solve the crime. Thus, these small clues may or may not qualify as evidence. So, further investigations need to be conducted on the potential evidence to decide who is more likely than not to have murdered the victim. In epistemology, unlike logic, we at no time seek 100% or absolute certainty. Accordingly, investigators will be trying to build a sturdy case to successfully declare someone guilty beyond a reasonable doubt. For that reason, they collect scattered bits of evidence and attempt to fit them well into a coherent story that makes the allegation that a certain person did the offense fit with the facts. Two things are noticeable regarding the method of the sleuth. One is that the case she constructs is accumulative. It is very rare for there to be a single part of the evidence that convicts the criminal. There commonly is a large number of segments of evidence: fingerprints, witnesses, receipts, and all kinds of things that helped to locate the suspect where the offense was committed. It is an accumulative case. Secondly, what the entire approach points out is something that we can call a fitting-ness way of dealing with issues: how does one well fit all these bits of evidence together into a cohesive story to present it as a good argument?

This may well be what transpired in most of the thought experiments discussed in the first chapter.[20] If anything constitutes a paradigmatic use of this sense of argument, it is Jackson's Fred thought experiment. The argument takes the following simple form: physicalism holds that all mental states can be understood merely by means of physical terms. Fred's color experience shows that physicalism leaves out at least some mental states like 'qualia' or 'raw feels.' For *having all the physical information about him won't help us to know everything about his color experience.* Therefore, physicalism is an unattractive view. In the same vein, in Jackson's Mary example, the argument goes as follows. Physicalism holds that everything can be understood merely by means of physical terms. Mary, who learned everything about the physical world in a black-and-white room, will learn about other colors existing in the world once released from that room. This shows that *Mary's previous physical knowledge is not enough to account for all that she knows.* Therefore, the thesis of physicalism is mistaken.

[20] In what follows, I will reconstruct those thought experiments as different kinds of epistemic arguments italicizing what was identified as intuition about each case.

Furthermore, functionalism or computationalism holds that the human brain or mind is computer-like with an information-processing system that operates on formal symbols. The Chinese room thought experiment shows that both the computer and mind fulfill computational operations. But *computers lack the cognitive states that the mind has*. Therefore, the mind is more than a computer program, and functionalism or computationalism about the mind is problematic. Likewise, the argumentative flow in Chalmers' Zombie case goes as follows. If physicalism, the view that physical stuff is enough to explain consciousness, is right, then there should be no possibility for there to be a world that is physically identical to but lack some components (e.g. consciousness) of ours. But *zombies, which are creatures that are physically our duplicates but lack our conscious experiences, are conceivable or knowable*. Therefore, physicalism is false.

In the same way, we can read Cohen's Lottery cases as follows. If fallibilism is right, then induction is enough ground for obtaining true knowledge. Cohen's Lottery cases show that this is not always the case: in case 1, *the subject does not inductively know that she will lose*; in case 2&3, *the subject inductively knows that she will lose*. So, fallibilism is not suitable, and induction is not enough ground for obtaining true knowledge.

Goldman's Fake Barn argument goes as follows. One knows that something is the case if and only if it is not accidental that one is right about its being the case. The subject of the thought experiment *Henry knows in the first scenario but not in the second*. Therefore, what determines whether our assessment is true or false is the distinguishability or discriminability of the situation. In addition, in Gettier's Smith-Jones ten coins case, the example is meant to work as an epistemic defeater of the evidentiary status of the concept of knowledge as justified true belief. The structure of the argument goes as follows. The boss told Smith that Jones would be offered the job, and Smith himself had already checked and got to know that Jones had ten coins in his pocket. Smith has a strong and true justification for believing that the one who will get the job has ten coins in his pocket. Smith, in the end, was offered the job, and it was unknown to him that he himself had ten coins in his pocket; hence *Smith does not know the proposition 'the one who will get the job has ten coins in his pocket' is true*. Therefore, the concept of knowledge as justified true belief is undermined.

Moreover, attacking descriptivism, according to which a proper name with a definite description associated with it refers to an object of the world, Kripke's Gödel–Schmidt case proceeds as follows. The proper name "Gödel" is associated with the definite description as the author of the incompleteness theorems. Gödel stole the script of the theorem from its real author Schmidt and published it in his name. But *we do not refer to Schmidt when we use the name, Gödel*. Therefore, descriptivism is mistaken.

Additionally, in Norman the clairvoyant case, BonJour criticizes the externalist version of foundationalism which holds that justified belief is an instance of knowledge even if the believer does not know what justifies it. Here is the argument. For a belief to be justified, the believer should know what justifies it. The subject of the thought experiment, *Norman, believes that the President is in New York City though he has no evidence either for or against this belief* (i.e., he does not know what justifies it). Therefore, Norman's belief is not an instance of knowledge, and the externalist version of foundationalism is false.

Another kind of argument is the one commonly referred to as an inductive argument (Juthe 2005). An inductive argument consists of inductively strong premises so that the conclusion is likely correct. An example is an argument of the form: 95% of the people in New York are democrats. Jones is from New York. Therefore, Jones is a Democrat. As we can read from the example, the premises of an inductive argument can be justified a posteriori; the argument flows from and through what is particular to what is general, and the premises of a correct inductive argument give rise only to a probable conclusion.

This may well be what transpired in the Truetemp case in which Lehrer argues against the externalist form of reliabilism, the view that the belief-world connection is sufficient to obtain reliable knowledge regardless of how this connection is made. Here, to remind the reader is the argument setup. The head of the protagonist of the thought experiment, Mr. Truetemp, head has been implanted by an unusual but reliable device. The device acts as a sensor to transmit information about the temperature to the computational system in the brain. Mr. Truetemp can tell what the temperature of any place is by like that. *But so, the intuition goes, Mr. Truetemp does not know what the temperature is when he truly says what it is*. Therefore, the externalist form of reliabilism appears to be woefully misguided.

A third and last category of arguments that have been given attention is called a conductive argument (Govier 2010). In a conductive argument, the premises do not need to be linked together; rather, they individually offer support for the conclusion. The support provided by the premises need not be considered collectively in determining the evaluation of the conclusion. Like in a standard inductive argument, where we cannot be provided with a *certain* conclusion but only a probable one, a conductive argument provides us only a probable conclusion; yet, it is considered a good argument. The difference between inductive and conductive arguments is that in the former, the premises collectively support the conclusion, while in the latter, the premises individually determine the evaluation of the conclusion. Here is an example of a conductive argument. We have no clouds in the sky. The weather forecast channel reported that it would be sunny in the coming 2 weeks. Therefore, rain most likely won't be seen tomorrow.

This may be what happened in the Charles experiment which (Nichols et al. 2003) called the 'Truetemp case'. In this experiment, the argument runs as follows. 32% of Westerns intuited that Charles knew. 12% of East Asians intuited that Charles knew. Therefore, folk epistemic intuitions most probably are not universal. Or, 68% of Westerns intuited that Charles did not know. 88% of East Asians intuited that Charles did not know. Therefore, folk epistemology most likely is not universal.

If the above argumentation is true, then we end up with two points: (1) thought experiments are structured as arguments and (2) intuitions are essential elements of the very arguments (thought experiments) they are parts of. If so, if (1) and (2) are true, then we are required to have more specific queries in mind. It seems that we have frittered away hours of valuable time asking and trying to answer a trivial question which could easily have been settled by changing its subject. That is, the question of whether intuition qualifies as an evidential source of knowledge was unsuitable, and it is here where we need to stop posing it. This evaluation question should have, instead, been asked but on the argument of which intuition was a part. If this is the case, I still have to answer some further questions in that domain: what and how is an argument capable of having a label for being denied or accepted? Bearing in mind the above distinction between the nature and structure of logical and epistemic arguments, they also should be evaluated differently. In what follows, I will try to clarify this difference in evaluation.

3.5.3 Evaluating Logical Argument

Like other evaluations, a logical appraisal requires standards an argument needs to meet in order to be worth commending and thereby not condemning. (Sinnott-Armstrong and Fogelin 2013) determine the main standard for assessing arguments, namely, validity with two subsidiaries: truth and soundness. An argument is said to be valid if and only if the content of the conclusion follows from the content of the premises. Otherwise, it is said to be invalid. That is, for an argument to be valid or consistent, the truth of its conclusion follows from the truth of all of its premises. The truth of a logical statement or proposition depends on its reliability. In Schechter's words:

> We can understand the claim that we are reliable about logic as the following thesis: the logical propositions we believe (upon reflection and discussion) are by-and-large true and the logical propositions we disbelieve (upon reflection and discussion) are by-and-large false (Schechter 2010, p. 438).

We infer this reliability from the consistency of the stated rule or measure. Ennis writes:

> In the area of internal consistency of tests, I noted the use of the word "reliability" to refer to consistency of repeated measure and the use of internal consistency as an indicator of this "reliability" (Ennis 1984, p. 8).

Pushing the discussion back to validity, just consider the following argument that passes the test for validity. All philosophers are thinkers. Socrates is a philosopher. Therefore, Socrates is a thinker. This is a valid argument because if the two premises are true, then there is no way for the conclusion to be false. Contrast this argument with the following example. If it is raining, then the ground is wet. The ground is wet. Therefore, it is raining. This argument is invalid because the truth of the two premises does not entail the truth of the conclusion. The ground could be wet because of different reasons other than rain. The ground could be wet because the sprinkler system could be on, for example.

But an argument might be valid even when some (or all) of its premises are false. For example, All fathers are male. Wisam is a father. Wisam is male. This argument meets the standard of validity: if the premises are true, it follows that the conclusion is also true. Nevertheless, the argument might not be true. In (Strawson 1952)'s words, it does not correspond with the facts. Suppose that Wisam is a mother. She has this name because she is Syrian, and in Syria, the name Wisam can be given to both males and females. Suppose further that Wisam, due to some abnormal hormonal activity, has all the physical qualities commonly associated with men, such as facial hair and small breasts. In this case, the second premise is false and accordingly, the argument is not free from defects. What all this means is that validity and truth are not the same things; they are two different things. It also means that for an argument to be good, all of its premises need to be true. Moreover, when an argument is valid, and its premises are true, we say it is sound. If it does not meet one or both of these standards, it is said to be unsound.

So far, I have been basically concerned with deductive arguments that are intended to meet the standards of validity. However, many arguments do not aim at satisfying these standards. Inductive arguments, for example, are intended to meet the standard of strength. Such arguments are planned to be appraised as strong or weak in accordance with the strength or weakness their premises offer to the conclusions. Here is an example of an inductive argument. All observed swans are white. The bird on top of my friend's roof is a swan. Then, most likely, it is white. This is a strong inductive argument. But the swan which has not yet been observed might be black. In this case, the argument will be weaker than before, yet it still has some force. The degree of its strength is less, but its premises can still provide support for the conclusion. In view of that, an inductive argument can offer only defeasible, partial, and gradual clues to the conclusion.

Last but not least, to evaluate a conductive argument, we have to assess the cogency of its premises (Govier 2010). A cogent argument means that the premises are acceptable (i.e., there is no available body of fact or information that they are false), they are relevant to the conclusion (i.e., both the premises and conclusion hang together well), and they together pave the way for the conclusion. Here is an example. Mary and John kissed each other. They celebrated Valentine's together. Therefore, they love each other. In light of the standard of cogency, this argument is cogent; because the premises are acceptable. Also, they are relevant to and collectively lead to the conclusion. If they were otherwise, this argument would not be cogent.

Having gone through presented how logical evaluation can be done and what are its appraising words, I shall be attempting, in what follows, to show how epistemic evaluation can be done and how that is possible in epistemically appraising terms.

3.5.4 Evaluating Epistemic Argument

Epistemic evaluation of an argument does not have the same aims and hence is not subject to exactly the same standards as a valid, strong or cogent argument. An epistemological approach to argument evaluation already has some proponents in the work of (Goldman 2003; Lumer 1991; Biro and Siegel 1992, 1997; Hoffmann

2005; Pinto 2001, chaps. 2–3; Feldman 1994; Govier 1992; Blair and Johnson 1987). There are then precedents for an epistemic approach to argument evaluation. Following this growing literature in epistemology, I hope that the following comments will add some contribution to its further manifestation and support.

To my knowledge, epistemology is the area of philosophy that concerns itself with the issue of knowledge and its nature, source and limits. Also, epistemology is the study of those things closely related to knowledge, such as justification, justification's relation to knowledge, whether having justified belief is possible at all, whether one can have justified belief due to internal or external factors, the value of knowledge, and so on and so forth. From this, one can proceed to extrapolate that "what makes a good argument good is its suitability for producing justified belief in its conclusion by means of justified belief in its premises" (Goldman 2003, p. 54). In other words, in epistemic evaluation, an argument needs to meet an interesting type of standard that has been largely overlooked by practitioners of logic, namely the standard of justifiedness. If, in the logical appraisal, an argument needs to meet some or all of the following standards of validity, strength, or cogency in order to be worth commending and thereby not condemning, in the epistemic appraisal, an argument needs to meet the standard of justifiedness. In light of that, an argument is epistemically good if and only if it produces justified beliefs both in its conclusion as well as premises.

To see this, consider Gettier's Smith Jones ten coins example. First premise: the boss tells Smith that Jones will be offered the job, and Smith himself has already checked and got to know that Jones has ten coins in his pocket. Second premise: Smith has a good reason or justification for believing that the one who will get the job has ten coins in his pocket. Third premise: Smith, in the end, was offered the job, and it was unknown to him that he himself had ten coins in his pocket. Conclusion: Therefore, Smith does not know even though he has a justified true belief.

This is epistemically a justified argument because the first premise is a supplier of twofold justification: the boss's testimony and Smith's observation that Jones has ten coins in his pocket. That being so, the first premise provides a ready supply of justification that makes Smith at least fairly well justified to proceed and infer the belief expressed in the second premise. As it happens, too, the third premise shows that this belief –although truly justified- is not correct as it was expected to be. For Smith was offered the job, and he himself has ten coins in his pocket. So, says Gettier, Smith, by random chance, got to know that the guy who will acquire the job has ten coins in his pocket. And that is enough justification that justified true belief is not sufficient to explain knowledge. It turns out that Gettier's conclusion that the concept of knowledge as justified true belief is undermined is justified. Again, the argument as a whole is justified because it produced justified belief in its conclusion by means of justified beliefs in its premises.

This is true for all thought experiments considered in the first chapter. Let us take a look at one more example, namely, BonJour's clairvoyant Norman, which is epistemically a justified argument, because the first premise, namely, "for a belief to be justified, the believer should know what justifies it," provides a conceptual framework of justification that in turn shows why the second premise "*the subject of the thought experiment Norman believes that the President is in New York City though he has no evidence either for or against this belief (i.e. he does not know what justifies it)"* is not justified. Together, these two premises are enough justification for BonJour's conclusion expressed in the third premise that Norman's belief expressed in the second premise is not an instance of knowledge, and the externalist version of foundationalism is void.

After all, epistemologists study arguments by studying not only argument forms but rather the content of the argument plays a significant role in the process of evaluation. That is the reason why two arguments may have the same form but completely contradicting content and different epistemic values, as in Cohen's Lottery Cases and Goldman's Fake Barn case. However, while logicians try to get validity and certainty, the goodness of any epistemic argument is cumulative: the whole approach is something that we could call a fitting-ness approach. That is, how do we well fit all these pieces of evidence together into one story to support its justifiedness? Hence, as a piece of evidence in an argument, intuition has epistemic value as long as it fits well with other pieces into one story to support its justifiedness.

If so, then it is natural to think that if an argument is of a type that can be based on epistemic reasons provided both by its premises and conclusion, that makes it apt for epistemic evaluation in terms of justification. The proposal here is construing the epistemic nature of intuitions as externalistic. The epistemic role of intuitions depends on the contingent facts of their aptness for making arguments apt for epistemic evaluation as justified. In this way, I am denying that the mental states of intuitions have epistemic status in any intrinsic way, by one merely having them or in their own rights. Intuition does not stand as evidence alone; it is part of a whole, where the main issue is whether this whole has enough parts that justify taking a stand on any issue.

However, to bolster the distinction between logical-epistemic evaluation and appraising terms, it is useful here to recall that Strawson has drawn a distinction between different types of appraisals. When one says that an object is beautiful, one is appraising it aesthetically. When one says that a person is good, one is appraising him ethically. In the same way, when we say that one's statements are consistent or inconsistent, one's argument is valid or invalid, etc., we are appraising the person's statements logically (Strawson 1952). Following this line, I have argued so far that when one says that an argument is justified or unjustified, one is appraising it epistemically. There are various kinds of appraisals; the epistemic appraisal is only one of many different (but legitimate) models.

Given the fact that validity is the logical value of argument depending on the truth of its premises and conclusions as being a reliable indicator of their consistency, contemporary epistemologists, I think, have confused logical appraisal with epistemic one while arguing about the issue of reliability or unreliability of intuitions; and this has confused the issue rather than illuminating it. Validity (in case of argument), truth (in case of premises and conclusion), and soundness (in case of both valid argument and true premises and conclusion) are the main values logicians use in their evaluation. They are the essential objects of logical appraisal, where we declare a man's remarks to be consistent or inconsistent. We contradict him, and, in doing so, we make an appeal to a counter-assertion of our own about the subject of his remarks (Strawson 1952, P. 1–5). If so, if validity, truth, reliability and consistency come under the category of logical standards and keeping in mind that the three positions on the epistemic status of intuitions considered in this chapter focus mainly on their reliability, then they just use the wrong evaluation standard to evaluate intuitions.

On the other hand, this is unlikely to be an adequate strategy for the purpose of epistemic evaluation. The epistemologist considers all competing and incompatible arguments, questioning them depending on whether or not a theory offers a good ground for plausible justifications in favor of it. More precisely, the criteria for the evaluation of epistemic arguments are different from those of logic. Unless epistemologists recognize the uniqueness of epistemic criteria, the epistemic appraisal words will continue to be used as if they were words of logical appraisal.

However, what if justifications conflict among epistemic peers? What if, for example, two epistemic peers epistemically evaluating an epistemic argument do so differently: one says it is epistemically justified, and the other says it is not? I will try to address this question in the next section.

3.5.5 *Justifiedness and the Challenge of Disagreement*

We, more often than not, disagree with others in philosophy. For that reason, we cannot be confident in the justification of an arguer's belief when we realize that everyone is an epistemic peer with conflicting evidence. In other words, suppose that it is true that an argument is epistemically good because it produces justified beliefs in its conclusion as well as premises. And suppose further that the evidence that justifies those beliefs is controversial and polemically gives rise to disagreement among epistemic peers. In such a case, it is not clear what argument is epistemically good and what is not. The evaluative question arises whether we should continue considering it a good argument or shall we discard it. This would, therefore, keep issues such as the epistemic standard of justifications in question since we do not know whether or not the relevant argument has a certain property that is epistemically worth commending or not commending. However, epistemic disagreement is one of the central ideas around which recent epistemic thinking revolves. There are, however, two main positions on this issue found in the recent literature. Here are their basic ideas.

On the one hand, some argue that disagreement is not possible and, accordingly, it has no effect on the evidence that justifies beliefs. This view is called the uniqueness thesis, which holds that "there is a unique rational response to any particular body of evidence" (Kopec and Titelbaum 2016, p. 189). That is, the same particular evidence does not, on its own, in fact, entail divergent justified appraisals amongst epistemic peers. To put it simply, E (evidence) on its own justifies B (belief). Roughly speaking, "it cannot be that epistemic peers who have shared their evidence can reasonably come to different conclusions" (Feldman 2011, p. 156). The camp of impermissivists or the authors who are sympathetic to the uniqueness thesis includes (Feldman 2006, 2007; Christensen 2007, 2009; White 2005, 2007, 2013; Matheson 2011, 2015; Horowitz 2014; and many more others).

On the other hand, permissivists or authors who are sympathetic to the anti-uniqueness thesis hold that "given the same total evidence, two rational subjects can hold different views" (Dogramaci and Horowitz 2016, p. 130). That is, disagreement is of great epistemic weight. For given the subjective, instrumental, relative, social, cultural, etc. factors that affect the relevant evidence, the same particular evidence does, in fact, give rise to divergent justified appraisals amongst epistemic peers. For example, religious disagreements have their consequence concerning epistemic obligations: when two epistemic peers with very different religious views dispute some matter, "the parties enter the discussion with different starting points that affect their judgment of what is evidentially significant and how it should be weighed" (Holley 2013, p. 36). Among authors who support the negation of the uniqueness thesis are (Rosen 2001; Douven 2009; Titelbaum 2010; Ballantyne and Coffman 2011; Brueckner and Bundy 2012; Decker 2012; Rosa 2012; Kelly 2010, 2013; Meacham 2014; Peels and Booth 2014; Schoenfield 2014; Kopec 2015; Weber 2017; Barz 2019).

Having mentioned the two main positions on the topic of epistemic disagreement, I am required, in what follows, to try to resolve the dilemma. To respond to the challenge from variation in justifications, one may, I submit, distinguish between two types of justification: local and global.[21] An argument is locally justified if beliefs arise in its conclusion and premises are justified by local evidence or local operative settings such as subjective, instrumental, relative, social, cultural, semantic etc. factors that affect the relevant evidence which accordingly gives rise to divergent justified appraisals amongst epistemic peers.

Consider Goldman's Fake Barn case. As mentioned above, the thought experiment was utilized to address the question of how the variability in our intuitive assessments can be explained. We must keep reminding ourselves that the argument runs as follows. One's knowledge that something is the case should not be at all accidental. We assess that Henry knows in the first scenario but not in the second. Therefore, the condition-based value of our intuitive assessments is determinable depending on the distinguishability or discriminability of the situation. Suppose a Kripkean like (Mizrahi 2016) reads this thought experiment; he won't agree that the concept of knowledge is not met. He will argue that Henry uses the word 'barn' in terms of the speaker's reference (i.e., what to the speaker Henry looks like a barn, which could have easily been a barn façade) and not in terms of semantic reference (i.e., the one real barn). Accordingly, Mizrahi will argue that, on the basis of these details, if we are asked, we will intuitively say that Henry knows what he perceived in both scenarios. And the Fake Barn case failed to show that there is a little bit of justifiedness in our intuitive assessments. If this is correct, then due to subjective,

[21] It is worth mentioning that (Goldman 1997) offers different classifications of justification. For him, there are social, personal and interpersonal justifications. He argues that a good epistemic argument must be guided by interpersonal justification.

instrumental, semantic etc., factors, the same body of evidence gives rise to divergent justifications amongst at least two epistemic peers, as in the case of Goldman and Mizrahi.

An epistemic argument is globally justified if the conclusion and premises are justified by widely shared evidence which accordingly gives rise to almost similar justification amongst epistemic peers. By widely shared evidence, I do not mean that there should be almost complete unanimity on the evidence at hand. Nor do I mean that there should be agreed by the majority of people on the evidence at hand. What I mean is that regardless of any unanimity, an epistemic argument is globally justified if beliefs arising in its conclusion and premises are justified by good evidence from an epistemic point of view.

If so, we are required to tell what this particular epistemic point of view could be. It is defined by the aim of excluding coincidentally true belief from knowledge. That is to say; it helps us to maximize rightfulness and minimize fallaciousness in beliefs, as in Gettier's original counterexamples where every premise plays the role of justification maker for the next premise. For these reasons and others, I take epistemic evidence to be a mental state relevant to justified belief. We determine it by considering the role or function it plays in argument, i.e., having an epistemic reason(s) that enables it to justify the belief(s) arising in the next premise. Building off those insights, I submit that an argument that has such kind of evidence is epistemically justified. The crucial point is that an argument that has justified beliefs in its conclusion and premises, which are justified by local evidence(s) is justified subjectively, instrumentally, semantically, socially, etc. but not epistemically.

3.6 Concluding Remarks

In this chapter, I have developed an account of the epistemic status of intuitions. By showing the shortcomings of the arguments of reliablists, skeptics, and perspectival relativists regarding the epistemic status of intuitions, I have shown that the evidential status of intuition does not depend on it itself, rather it rests on its being combined with other pieces to make up a whole —thought experiment or argument. The main failure of the aforesaid pros and cons accounts of the evidential status of intuition is that they evaluate intuition separately. Alternatively, in my view, intuition has evidential status if and only if it plays a role in making up an epistemically justified argument.

I think that my picture of the nature of epistemic intuitions and their evidential role can be summarized as follow. Intuition is an intentional mental state. But it is non-propositional, non-phenomenal, and individuated by its origin, i.e., thought experiment. And such intuition transcends the traditional a priori-a posteriori distinction. Furthermore, epistemic intuition arises from an epistemic thought experiment, which, in turn, is structured as an epistemic argument that is, in nature, different from a logical one. There is an important and crucial distinction between epistemic and logical evaluation. For that reason, the epistemic argument needs to be evaluated by epistemic standards and not logical ones. If so, the question of epistemic status does not apply to intuition separately. Rather, it is applicable to the argument itself, which can be either epistemically justified or unjustified or non-epistemically justified or unjustified. When it is epistemically justified, it is so by virtue of widely shared evidence(s). Given the fact that it is a mental state with an epistemic reason which enables it to play the role of evidence, intuition has epistemic weight, depending on how much it helps us to build the epistemic justifiedness of the argument.

Now that we have articulated an account of both the nature of epistemic intuitions and their evidential status, we are in a position to assess the immunity of its arguments. One way to do this is to show that it does not fall victim to such arguments offered by 'intuition-deniers' who—like Timothy Williamson, Max Deutsch, and Herman Cappelen —typically direct their attention to rejecting the claim that epistemologists accept intuition-based epistemology, especially at face value.[22] It is to that task I now turn.

[22] Those philosophers were called 'intuition-deniers' by (Nado 2016).

Chapter 4
Epistemic Intuition in Light of Intuition-Deniers

4.1 Introduction

As noted in the previous chapter, several views about intuition – reliabilism, skepticism and perspectival relativism – hold that intuitions play a central role in philosophical theorizing.[1] Reliabilism believes that these intuitions function directly as evidence in philosophical arguments. Not less importantly, skepticism asserts that intuitions are usually thought to play such a role, although this is an inappropriate role for intuitions to play. And perspectival relativism admits that intuitions can be taken as a relative source of evidence. Accordingly, it seems as if contemporary analytic philosophers often use intuition as evidence in their philosophizing. I will refer to this sort of thinking as centrality.[2] Contrary to that, several philosophers have recently raised different objections to this thesis of intuition-based evidence denying the claim that philosophers centrally depend on intuitions as evidence. Thus, intuition-deniers present a serious challenge to the claim of intuition-based evidence in philosophy.[3]

The main goal of this last chapter is to assess the quality of the account of the nature, and epistemic role of intuition defended in the previous chapters in light of some of the main current objections that have denied its evidential role. Since the literature on this topic is huge and has lately been increasing at an exponential rate, I will only attempt to look at a few key objections to intuitions as evidence.[4]

---

[1] That is not to say that those philosophers are saying that intuition serves as playing the only substantial or even the most evidentially vital role in philosophy. They only claim that intuition has a central evidential role to play in philosophy without denying that there may be some other entity which equally plays a centric role.

[2] I borrow the term "centrality" from (Cappelen 2012).

[3] I will refer to this sort of objection as the "anti-centrality" response.

[4] Main proponents of the theses of intuition-deniers include (e.g., Gendler 2007; Earlenbaugh and Molyneux 2009, 2010; Molyneux 2014; Ichikawa and Jarvis 2013).

M. Hamdo, *Epistemic Thought Experiments and Intuitions*, Philosophical Studies Series 150, https://doi.org/10.1007/978-3-031-33480-1_4

Specifically; I will focus on the following views: the thesis of anti-centrality given by Herman Cappelen, the philosophical method of using counterfactual conditionals given by Timothy Williamson, and Max Deutsch's view of anti-intuition-based methods of analytic philosophy.

While there are crucial differences between Williamson, Deutsch and Cappelen, all of them are inclined to adopt the extreme view that intuition never plays any evidential role in philosophy. More precisely, they mainly argue (1) against psychologizing the evidence in philosophy: intuition is a psychological event, and by considering it as evidence in philosophy, one psychologizes the idea of philosophical evidence, (2) due to the disagreement among philosophers at the level of the concept of intuition, they also express skepticism toward whether there is a referent of the term 'intuition' at all and (3) they attempt to prove that philosophers develop views not by means of intuition, but via argument.

In this chapter, I shall defend centrality by discussing these three lines of argument and responding to them, showing that they fail to affect the account of intuition I attempted to establish in the previous chapters. Before proceeding, it may be useful to explain centrality clearly. It is a descriptive claim about the role most meta-philosophers give to intuition. It is not a normative claim implying that intuition must play its present role. However, it describes the use of intuition in philosophy. It does not describe philosophers as thinking of themselves as granting intuition a crucial evidential role.

This chapter will be divided up into five parts. In Sect. 4.2, I present and discuss Cappelen's, Williamson's, and Deutsch's intuition-rejecting accounts. In Sect. 4.3, I describe the main three intuition-rejecting arguments the above-mentioned accounts consist of. In Sect. 4.4, I show how these arguments fail to have any effect on my account. And then, I close this chapter with some concluding remarks.

4.2 Revisiting the Accounts of Intuition-Deniers

Before I turn to look at each of their arguments, however, I now briefly revisit each of the accounts of intuition-deniers to restate their tasks and goals as conceived and illustrated in their own texts. Otherwise, the cart will seem to be as if it is put before the horse.

4.2.1 Cappelen's Thesis of Anti-centrality

Given the epistemologists' various positions and arguments for and/or against intuitions discussed in the previous chapters, most philosophers have been encouraged to believe the thesis of centrality by which Cappelen intends to mean "the claim that contemporary analytic philosophers rely extensively on intuitions as evidence" (Cappelen 2012, p. 1). Cappelen questions this thesis by posing a humble supervisory question: is it true that contemporary analytic philosophers rely on intuitions as evidence? (Cappelen 2014b). The answer, he argues, is that centrality is simply a false descriptive claim about what philosophers do while philosophizing (Cappelen and Winblad 1999; Cappelen 2014a). Instead, as Cappelen observes, it is not true that "philosophers in general rely on intuitions as evidence when they do philosophy" (Cappelen 2012, p. 3). In fact, this is what he suggests as the thesis of anti-centrality. However, Cappelen attempts to refute centrality by evaluating and then undermining two main arguments given by proponents of centrality in favor of it: the argument from intuition-talk and the argument from philosophical practice.

According to the argument from intuition-talk, many contemporary philosophers engage in intuition-talk in the way that such talk is crucial in serving as evidence to their philosophizing. Taking those philosophers at their word, Cappelen argues that in English as well as in contemporary philosophy, there is a variety of uses of intuition-terminology and cognate terms but in a promiscuous way. That is to say that intuition-terminology "is used, but with a sense that doesn't support centrality, i.e., it isn't used to denote a kind of judgment or mental state that serves as evidence or a source of evidence)" (Cappelen 2012, p. 25). This indiscriminate usage endorses the claim that philosophers do rely on intuition as evidence while theorizing.

Cappelen starts from the observation that there is a sharp difference in the use of intuition-terminology between ordinary speakers and intuition-theorists. According to him, the ordinary usage of intuition-terminology is quite thin. For example, in order to understand the relevant features of the noun 'intuition', he begins to inquire what things the adjective 'intuitive', the adverb 'intuitively', and the word 'seem' are applicable to in ordinary English. He finds that, in ordinary English, the adjective 'intuitive' and the adverb 'intuitively' can be applied to a variety of different kinds of acts and entities such as gadgets, melodies, singing, operating systems, Basketball video game passing, ways of stating arguments or points, connections with people, chess playing, dance partnerships, speech acts, propositions, ways of knowing, remarks, philosophy and claims. As things stand, "there is one feature that stands out when these cases are considered: there is some kind of ease, effortlessness, or spontaneity involved" (Cappelen 2012, p. 33). This is not to say that the relevant things can be performed without relying on immediate perception (e.g., dancing), chains of reasoning (e.g., playing chess) or memory (e.g., singing).

Cappelen also makes two other interesting observations: first, what those acts and entities inform us is that the ordinary usage of 'intuitive' and 'intuitively' is context-sensitive. In particular, 'intuitive' is gradable in the sense that a proposition can be intuitive to a certain degree. As a gradable adjective, according to Cappelen, 'intuitive' is context-sensitive because whether it is true of something depends on the comparison class, which is fixed by the context. Second, ordinary speakers often use 'intuitively' and cognate terms to express a hedged attitude towards a judgment. By the term 'hedge,' it is meant "an expression that functions, at least in part, to weaken the speaker's commitment to the embedded sentence" (Cappelen 2012, p. 36). They frequently say 'intuitively p' in cases where they are unsure of the judgment that p, flagging a weaker commitment to their claim. Uses of 'intuitively' and cognate terms, such as ‘apparently’, ‘seems’, ‘believe’ and ‘think’ are examples of those hedges.

However, the noun ‘intuition’ is used in the same way the adverbial and adjectival forms are utilized. To show that this is the case, Cappelen discusses many straightforward examples restricting his “attention to cases where the object of intuition is something propositional and truth-evaluable ... and the act in question is a judgment” (Cappelen 2012, p. 39). In light of this, the content of what is assumed in non-philosophical and non-scientific literature as the ‘faculty of intuition’ or the ‘sixth sense’ is just ‘BLAH’ and not evidence or source of evidence. Proponents of centrality may say that its support comes from the use of terms such as ‘seem’. Giving many examples, Cappelen responds to those who equate ‘seem’ and ‘intuitive’ that ‘seem’ is often used in ordinary English to function as a hedge term, and it is not universally interchangeable with ‘intuitive’.

On the other hand, most theorists of intuition, according to Cappelen, regard intuition as a mental state with at least one of the following three features. First, intuitions have a special phenomenology. Second, intuitions are based on nothing but the conceptual competence of the subject. Third, intuitions justify, but they themselves don’t need further justification.

After noting these differences between intuition-theorists and ordinary speakers, Cappelen proceeds to discuss most philosophers’ intuition-talk. On his view, the claim of centrality proponents that philosophers do not use intuition-terminology as it is often defectively used in colloquial English, but rather, they constructively use it in a theoretical or technical way in which it refers to a mental state with any of the above three features playing the role of evidence seems far-fetched, simply because:

> there is no agreed upon definition of ‘intuition’. There are no agreed upon paradigms. There is minimal unity in usage between different schools and subdisciplines and there is no group of experts within the discipline who agree on how the term should be used (Cappelen 2012, p. 52).

It is difficult to define ‘intuition’ in an uncontroversial way among intuition-theorists even. Accounts vary and conflict: many intuition-theorists, such as Bealer, take ‘intuition’ to indicate a *sui generis* mental state. He claims that intuition is an *intellectual* seeming, irreducible, natural and propositional attitude which occurs episodically in phenomenological terms. Others, such as Chudnoff, take ‘intuition’ to be a form of intellectual perception. It is a phenomenal, *sui generis* mental state or experience which is both constituted by other experiences, e.g., conscious thoughts, imaginings, etc., and accompanied by a special phenomenology. Some intuition-theorists, such as Sosa, think that intuition is a rationally and intellectually competent seeming that attracts to assent a propositional content based only on consciously understanding it. In addition, some intuition-theorists, such as, e.g., (Goldman and Pust 1998, Sosa 2007b; and Ludwig 2010), think that intuitions or intuitive judgments must be built only on conceptual competencies.

Cappelen takes all this to be instances of a view according to which contemporary philosophers do not agree on what ‘intuition’ indicates. That being so, ‘intuition’ might be “a device of hedging: a way of qualifying a speech act (much like ‘I think' functions in some utterances of 'She's in Paris, I think'). So used, it does not denote any kind of mental state or source of evidence." Or it might be a context-sensitive term "both in English and as used by philosophers." Some intuition-theorists argue that "if 'intuition' denotes a mental state, it is non-factive: you can have an intuition that p, even if p is false. It is also typically assumed that an agent can have the intuition that p even if she does not believe that p" (Cappelen 2012, p. 11–12).

With reference to some quotations taken from a number of great philosophical texts, Cappelen outlines three hedge-strategies that guide most cases of the passages in which philosophers use intuition-vocabulary. The first strategy is simple removal, according to which intuition and related terminology are meaningless to the extent that one, in a very wide range of cases, can remove them from the text with "no effect on the substance of the argument, but clarity and argumentative rigor will be improved (Cappelen 2012, p. 61). The second strategy is snap or unreflective reasoning. In many cases, simple removal does not work. Making a quick browse through philosophers' use of intuition-terminology and cognate terms, he explains that, in such cases, intuition and related terminology can be reinterpreted as a snap or unreflective reasoning. The third strategy is pre-theoretic commitment or meaning unaided by investigations, according to which philosophers' use of intuition-vocabularies accords with our pre-theoretic ideas.

Keeping those three strategies (simple removal, snap, and pre-theoretic) in mind, Cappelen concludes that the only reinterpretation of intuition and related terms is a tool of hedging. Yet, this is not to say that these strategies capture all philosophers' use of intuition-vocabulary. Some philosophers' use of intuition-terminology cannot be interpreted even as a hedge use. It is not easy to interpret philosophers' uses of intuition-vocabulary by these three strategies, as in what Kripke means by 'intuition' and cognate terms in some passages of his book *Naming and Necessity*. Cappelen writes, in some cases, "it is massively underdetermined what exactly is contributed by 'intuitively' or 'intuition'-vocabulary (Cappelen 2012, p. 77).

Moreover, the above strategies make no reference to the three elements mostly attributed to intuition: a special feeling or phenomenology, conceptual competence or truth, and special justificatory status. Thus, these interpretative strategies are not in support of centrality, and thereby the strategy called as 'explaining away intuitions' is clearly on the wrong path.

The upshot of the previous discussion was that the strategy of 'explaining away intuitions' is not a charitable interpretation of philosophers' use of intuition vocabularies. Let us turn to the argument from philosophical practice.

The Argument from Philosophical Practice, on the other hand, focuses primarily on what philosophical practice is like and how philosophers do their philosophical argumentations. Is it true that they prominently use the method of thought experiments relying on intuition as evidence for and/or against their philosophical views? In order to avoid any general answers to these questions, Cappelen empirically studies some thought experiments from contemporary philosophy to show what philosophers really do while philosophizing. To that end, he focuses first on "methodological rationalism," followed mainly by Bealer, BonJour, and Plantinga.

He shows how many claims from that tradition, which celebrates the usage of intuition as evidence in philosophy, are common in it and its opponents or experimentalists, who aim at empirically investigating intuitions used in philosophy. An example of these elements of the agreement are intuitions which are phenomenal mental states with a privileged epistemic status essentially used as one of the core methods in philosophy i.e., the method of cases (Cappelen 2012, p. 105–6). Cappelen argues, as we have seen above, that according to most intuition-theorists who support centrality, an intuition significantly has the following three features. First, it seems true that intuition has a special phenomenology. Second, intuition has a non-inferred, special, epistemic status, i.e., a recalcitrant, default, justificatory status (it justifies but needs no justification). And third, intuition is based solely on conceptual competence (Cappelen 2012, p. 111–14).

He looks at ten paradigmatic thought experiments to see whether we can detect evidence of the presence of those three diagnostics (Cappelen 2012, p. 130–87). He focuses on where centrality proponents usually claim that there is an appeal to intuitions and does not find any sign of reliance, not even ‘implicit’, on intuitions as evidence. Given the investigation of a number of paradigmatic thought experiments, he draws various conclusions from the study as such. The overall conclusion is that on any precise description of intuition and cognate terms in philosophical practice, there is no reliance on intuitions as evidence. Consequently, the argument from philosophical practice provides no support for centrality and thereby, centrality is false.

And then he considers a restricted version of centrality according to which a proponent of centrality usually assumes that intuition plays a central role in one or all of the following four practices: method of cases, armchair and a priori activities, conceptual analysis and rock-bottom as a starting point for arguments. But, (Cappelen 2012, p. 205–218) argues that these four practices can be explained in different ways. And since the considered thought experiments showed no defense of this claim, the role of those practices in philosophy may be just another false belief philosophers had about the place of intuitions in philosophical practice. Just as philosophers had a wrong belief that philosophers rely on intuition as evidence while theorizing, they had a false belief that intuition plays a central role in philosophical practices.

Having presented Cappelen’s main arguments, I would like to comment on them briefly. John Bengson (2014) responds to this refutation of Cappelen, stating that Cappelen’s problem lies in his assumption that either philosophers’ use of intuition-terminology is ordinary usage or it is technical usage. Instead, Bengson says philosophers’ use of intuition- terminology can be in-between, non-comparative, or discriminative usage. It denotes a conscious state or event that plays a substantial epistemic role. It is, therefore, centrality-friendly usage.

If so, if such use is there, then Cappelen’s view of Kripke’s use of ‘intuition’ as a mere pre-theoretical attitude without granting it any evidential weight is wrong. Citing Kripke’s intuition-talk in *Naming and Necessity*, Bengson insists that Kripke, at least at times, employs a centrality-friendly use of ‘intuition’, denoting something he relies on for positive epistemic status. And, since Cappelen himself says that Kripke’s use of ‘intuition’ influenced a generation of philosophers, then he should admit that their use of intuition is not simply a kind of intellectual or verbal virus or BLAH.

Bengson discusses two features out of the three listed and criticized by Cappelen as being habitually considered by philosophers as features of intuition. First, intuition always comes with a special phenomenology. Cappelen says one gets no special feeling or phenomenology while intuiting. It is just an anecdotal confirmation by some intuition-theorists. Bengson responds that whether or not intuition has its characteristic phenomenal properties is a further and irrelevant issue to the basic idea that philosophers rely on a conscious intellectual state or event for positive epistemic status. Bengson adds that Cappelen privileges the adjective 'intuitive' and adverb 'intuitively' while seeking to understand what was meant by the noun 'intuition', but it is the verb that anchors or adheres to the noun.

Second, justifies but needs no justification. This suggests that intuition has some fundamental rock-bottom justificatory status. Cappelen says philosophers rely on arguments, not on intuitions, to support their positions. In reply, Bengson produces his "reconciliation strategy" in that intuitions and arguments often interact in a friendly way. They need not be competitors because an argument may play one or the following roles: buttressing, capturing, and guiding the relevant intuition. And intuition may play one of the following roles: prompting further reflection and argument and problematizing or producing a controversy to be resolved or explained through further reflection and argument. Thus, the interaction between intuitions and arguments is complementary and not competitive.

Bengson concludes that Cappelen is right in rejecting the idea that intuitions have the function of confirming or disconfirming a claim or theory, but he is not right in rejecting this caricature on the basis of the idea that philosophers do not rely on intuitions in their thought experiments. This caricature should be rejected because intuitions play many more roles than such a caricature allows.

Here, one may ask, what are the many roles that intuitions play? In a footnote meant to identify those roles, Bengson mentions inference and dialectic. The problem with this is that it reduces the epistemic role of intuitions to dialectic. To be dialectically effective, an intuition about a subject matter does not need to be true. It just needs to be believed by some audience. Then this audience will consider it as a standard to subject other theories, beliefs, intuitions, etc., to it, as in the case of reflective equilibrium. We have already discussed the problems reflective equilibrium faces. However, some epistemic intuitions may be strong and widely shared but untrue, as in intuition: knowledge is a justified true belief. This intuitive judgment is no more comfortably true after Gettier-type counter cases.

In addition, in this sense, one may have a dialectic about knowledge or any other epistemic subject matter. Moreover, if an audience depends on its intuition to be justified upon a reason, and that reason depends for its justification upon another reason, and so on, then infinite regress seemingly ensues. On the other hand, to be epistemically effective, an epistemic intuition needs to be evidential. As we have already seen in the previous chapter, an epistemic intuition may have an evidential status depending on how much it helps us to build the justifiedness of the epistemic argument it arises from.

While Cappelen seeks to interpret both intuition-talk and philosophical practice in order to prove that they do not support centrality, Williamson provides an argument against centrality while examining how philosophers think in making a judgment.

4.2.2 *Williamson's View of Intuition as a Counterfactual Conditional*

Williamson raises the question of what underlies the thought experiment-induced judgment. Using the imaginary Gettier's ten coins case as canonical examples, (Williamson 2007a) describes his own positive conception of how to adequately understand thought experiments in philosophy.[5] To that end, he actually refutes the most widely accepted analysis, according to which the subject, in this case, has a justified true belief without knowledge and hence Gettier concludes that justified true belief is not sufficient for gaining knowledge.[6]

Simply put, Gettier's ten coins case starts from the idea that justified true belief is a sufficient condition for knowledge. They provide two counterexamples of this notion that makes use of two elementary principles: logical and epistemic. The logical principle: the logical consequence from a false premise cannot be true, and the epistemic principle: deduction is the right epistemic way of transmitting the justification from the premise to the conclusion of an argument. On this basis, Williamson argues, Gettier's ten coins case begins from the idea that justified true belief is a sufficient condition to constitute knowledge. Gettier's objection consists in the possibility for one to have a justified true belief which is not knowledge. According to Williamson, this analysis has many disadvantages. He states:

> One disadvantage of the abstract argument [i.e., the above analysis of Gettier-style thought experiment] is that it rests on several very general claims for which we might find adequate support hard to provide. In particular, it assumes that a belief essentially based on a false belief does not constitute knowledge. Can we take that for granted? (Williamson 2007a, P.181).

Williamson finds an answer to his own question as follows. Gettier's objection can be identified with the following possibility: if the Gettier case had happened, then the subject would have had a justified true belief without knowledge. Thus, the judgment made in the Gettier case has as its content the counterfactual conditional. Counterfactuals thus have a central position in his account of philosophical thought

---

[5] Note that by 'thought experiment,' he understands a form of argument which in turn is a source of philosophical knowledge. See (Williamson 2013b).

[6] When it comes to the concept of knowledge, Gettier presents a negative view of knowledge. He tells us what knowledge is not: knowledge is not a justified true belief. While (Williamson 2000) provides a positive view of knowledge. He tells us what knowledge is: knowledge is a kind of mental phase sensitized to the knower's environmental context within which she operates.

experiments. If so, Gettier can claim, Williamson thinks, that his case is possible, not that it is actual.

In addition, the usual description of Gettier's case is a very general picture, and no doubt omits potentially relevant details that need to be filled in. Williamson writes: "in philosophy, examples can almost never be described in complete detail" (Williamson 2007a, p. 185). Suppose, for example, one's true belief p prompts unpleasant memories. These memories will produce doubt about her false belief q. The result of these interfering factors, whether intended or unintended, may be losing justification for q rather than obtaining one for p. But if so, we cannot completely be assured that every Gettier-style thought experiment that satisfies the picture is one where there is justified true belief without knowledge. Williamson, therefore, thinks that our verdict is a counterfactual conditional: "if there were a Gettier case, that case would be one of justified true belief without knowledge" (Williamson 2007a, p. 185). The case goes from a modal premise (if the Gettier case had happened) to a modal conclusion (then the subject would have had a justified true belief without knowledge). In very rough terms, it is the counterfactual conditional, which allows people to fill in the details of the relevant thought experiment. That alone makes the thought experiment a significant and substantial philosophical method.

The results of a counterfactual conditional have been habitually expressed by what is called Gettier's intuition: "If a thinker were Gettier-related to a proposition, he/she would have justified true belief in it without knowledge" (Williamson 2007a, p. 195).[7] Hence, what epistemologists have generally termed 'intuition' is, in fact, a counterfactual conditional. Williamson states that the epistemology of counterfactuals in philosophy requires "no dedicated faculty of intuition" (Williamson 2007a, p. 178). The psychological basis of thought experiment judgments, according to Williamson, is far more complex than "the brute simplicity which the term 'intuition' may suggest," for one usually uses the term to "pick out a special psychological ... kind" (Williamson 2007a, p. 216).

As it stands, intuition is considered to play the role of evidence in contemporary analytic philosophy; Williamson writes: "intuitions are presented as our evidence in philosophy" (Williamson 2007a, p. 214). Yet, disagreement is the typical feature of epistemology when it comes to the question of the nature of this evidence; he states: "there is no agreed or even popular account of how intuition works, no accepted explanation of the hoped-for correlation between our having an intuition that P and its being true that P" (Williamson 2007a, p. 215). While intuition usually refers to a homogeneous psychological kind, philosophical judgments refer to various and heterogeneous psychological processes. Centrality, therefore, fails to capture the heterogeneous feature of philosophical judgment.

Epistemologists, according to Williamson, agree with the idea that intuition plays the role of evidence, but they do not agree on what is the nature of this

---

[7] It should be noted that intuition was not mentioned as the main motif of (Williamson 2007a). Rather, the main theme of the book is to uncover the "unexceptional feature of philosophy" (p. 4).

evidence. However, Williamson does not commit himself to any positive view of the nature of intuition. Williamson expresses skepticism about the idea that there is a special, authoritative, discriminable, or psychological faculty of intuition. Williamson argues against the idea that intuition is a key part of philosophical evidence.

In addition, Williamson not only has denied the psychological faculty of intuition, but has rejected to characterize it as having a certain kind of epistemological, phenomenological, perceptual or intellectual, conscious or unconscious, virtuous or vicious, a priori or a posteriori, external or internal, apparent or certain, probable or improbable, weak or strong, direct or indirect ability which provides non-inferential belief or inclination to believe with a positive epistemic status, common sense, what passes for it in one's culture, judgment attached to its truth conditions, or whatever can be called reliable evidence in philosophy.[8]

He thinks that intuition is derived from the capacity for judgment. When a judgment is saliently doubtful, we call it intuition: intuition is a doubtful judgment.[9] Consequently, Williamson warns that we cannot ground philosophical evidence in intuition and says: "philosophers might be better off not using the word "intuition" and its cognates" (Williamson 2007a, p. 220).[10] So conceived, the question arises is how are we to answer the questions about what should be called evidence in philosophy?

Straightforwardly, he thinks that evidence in philosophy has none of the above-mentioned characteristics philosophers use to ascribe to intuition.[11] Here is his criterion which evidence in philosophy must meet:

> Our evidence in philosophy consists of facts, most of them non-psychological, to which we have appropriate epistemic access. Consequently, there is a one-sided incompleteness to descriptions of philosophical methodology, and attempts to justify or criticize it on that basis, if formulated in terms neutral over the extent of that evidence. For instance, in describing some philosophers as believing or having the intuition that P, one fails to specify whether their evidence includes the fact that P (Williamson 2007a, p. 241).

Unlike other philosophers who claim that reflective equilibrium really yields an adequate method in philosophy and in epistemology, in particular, Williamson

[8] Since experimental philosophers' attacks are on intuitions of actual life thought experiments, they are included in this skepticism. See (Williamson 2016). However, experimental philosophy's data depends on intuitions of philosophically untrained subjects, the defenders of intuitions warn. Experimentalists reply that there is no evidence that the same story does not go for intuitions of philosophically expert subjects. (Williamson 2011) argues that experimentalists do not produce evidence that philosophical training is less effective than other cognitive tasks when it comes to intuitions.

[9] See (Williamson 2004).

[10] Notice the difference between Williamson's view and Cappelen's. Cappelen refuses the thesis of intuition centrality in philosophy: the claim that contemporary analytic philosophers rely on intuition as evidence. In comparison, Williamson doubts the existence of intuition at all. Williamson takes a more radical route.

[11] The knowledge of intuition is determinable by counterfactual conditionals. See (Williamson 2007b).

argues that reflective equilibrium, in fact, fails to provide a significantly proper method for an epistemological evaluation in philosophy. The reason why it fails to do so is that reflective equilibrium is an iterative process between a set of intuitions to bring them into harmony. So conceived, reflective equilibrium is analogous to "the mutual adjustment of theory and observation in natural science" (Williamson 2007a, p. 244). In philosophy, intuition is considered evidence, and it arises in a thought experiment. But, as we have already seen, Williamson's account of the thought experiment identifies it with the counterfactual conditional. This entails that reflective equilibrium fails to be the proper method in philosophy. Here is, for Williamson, the reason:

> since real life counterexamples [i.e. actual or real thought experiments] will sometimes do in place of imaginary ones, observed facts are sometimes relevant evidence. Talk of reflective equilibrium fails to address such issues (Williamson 2007a, p. 244).

That is to say that reflective equilibrium fails to have epistemic access to the observed facts that are sometimes considered to be relevant evidence. By depending on intuitions as evidence, reflective equilibrium may easily "slip into the illusion that our epistemic access to such psychological facts is unproblematic" (Williamson 2007a, p. 245). The danger, then, is that reflective equilibrium reconstructs the concept of philosophical or epistemic argument in purely psychological terms. We need to know, Williamson argues, that, in philosophical or epistemic argument, the facts which can play the role of evidence are the contestable facts. Its contestability, then, is what makes the philosophical or epistemic argument worth listening to. Thought experiments understood as counterfactuals can provide evidence as such.

Having presented Williamson's main arguments, I would like to briefly comment on them as follows. The main goal of Williamson's account just considered is to call into question the method of cases, the widely relied-upon method in contemporary analytic philosophy. This method is best known for its reliance on intuitions as a reliable source of evidence. As a matter of fact, Williamson's dependence on counterfactual conditionals to model thought experiments lack a generality requirement: it is an exclusive account, focusing only on the Gettier case. An opponent of Williamson's view might say that in order to get a deeper understanding of what thought experiments are, an account is required to be able to be applied to different types of thought experiments. Williamson's account does not meet this requirement. It is restricted to a modal thought experiment.

Contrary to Williamson's assumption, I have argued that an epistemic thought experiment is structured as an epistemic argument that has to be assessed in epistemic appraisal terms. I have also used many examples of thought experiments, showing what epistemic role intuition can play for a thought experiment to be epistemically successful. It is true that Williamson states that his view applies to different thought experiments. But this claim does not go beyond stating it. Perhaps, he is taking it for granted that once he states his account, there will be a consensus on the idea that thought experiments are counterfactuals. To my knowledge, this consensus has not been reached.

In any event, Williamson's target remains centrality, for he explicitly claims that his matter of interest concerns "how philosophy is actually done" (Williamson 2007a, p. 6). But Williamson's position seems to be more moderate. He is not saying that philosophical evidence does not consist of intuition. Instead, what he is saying is that philosophers should not use intuitions as evidence. However, there is still a crucial difference between Williamson and other critics of centrality, like Deutsch, who is inclined to adopt the extreme view that intuition never plays any evidential role in philosophy.

4.2.3 Deutsch's Debunking the Myth of Intuition

Max Deutsch examines and defends the methods of analytic philosophy against the challenge raised by the practitioners of the recent movement of experimental philosophy (xphi). Epistemologically speaking, the challenge concerns the 'myth' of the evidentiary role intuition was claimed to play in analytic philosophy. He writes: "philosophical arguments never appeal to the intuitiveness of a judgment about a case in order to justify belief in that judgment" (Deutsch 2015, p. 76–77). This 'myth' has been, Deutsch argues, around both experimentalists (xphiles) as well as Reliablists. Yet, the main target of Deutsch's criticism is xphi.

To that end, (Deutsch 2015) begins by drawing a distinction between 'negative' and 'positive' xphi. Negative xphiles, according to him, collect data on folks' intuitions, showing how these intuitions vary from people to people so as to question a particular argument, condemning the use of intuitions. On the other side, positive xphi calls certain arguments into question, but it declines to condemn the use of intuitions. Deutsch writes:

> But, while negative xphi involves collecting data about intuitions and "intuitional diversity" (that is, diversity in intuitions between different groups of people), this is really only a means to an end, which is both to call certain traditional philosophical arguments into question and, most importantly, to raise a worry about the epistemic value of the philosophical intuitions negative xphiles take to be involved in those arguments. There is a sense in which negative xphiles do not much care about who has which intuitions; they are mostly concerned with causing trouble for those who do care. Positive xphi, on the other hand, does not quite condemn the use of intuitions, though it can be quite critical of traditional philosophical arguments and conclusions, namely those that it takes to be dependent on explicit or implicit claims about who has which intuitions" (Deutsch 2015, pp. 1–2).[12]

Stating that his focus is mainly on the negative branch of xphi, he describes the findings of five current xphi studies showing how these findings either have been or

[12] Simply put, the distinction between the negative and the positive xphi, according to Deutsch, can be fleshed out as follows. Both negative and positive programs of xphi seek to determine people's intuitions by making use of the methods of polling, survey, etc., to show truth-irrelevant variability in intuitions. Also, both agree that philosophers use intuitions as evidence in philosophy. But while positive xphi does not take this variation as a reason to condemn the philosophers' use of intuitions as evidence in their philosophizing; negative xphi does.

might be utilized to call the traditional methods of analytic philosophy into question. To bolster his concentration on the negative xphi critique, he presents more studies arguing that negative xphiles' criticisms are philosophically not interesting, except what he calls pragmatic distortions, in which the contextual features of a question can cause the subjects of a survey to interpret the such question in one way rather than another.

However, Deutsch correctly observes that attempting to articulate what intuition is precisely is a problematic task due to the disagreement amongst analytic philosophers on the topic of the nature of intuition. Some define intuition in terms of allegedly its distinct characteristics like being spontaneous, non-perceptual, or non-inferential. Others just equate intuition with mere judgment, belief, or disposition to believe. The trouble, Deutsch argues, is that intuition is something that cannot be fully captured with a reasonably clear reference. That is why "for every proposed analysis of intuition, a host of problems arise, usually including a variety of intuitive counterexamples" (Deutsch 2015, p. 26). For, suppose you have adopted a theory according to which intuition is belief. And the examination showed that philosophers sometimes rely on intuition as intellectual seeming. In that case, it would be very plausible to think that your theory is false and that the thing you were relying on was not evidence.

In view of the fact of disagreement on what intuition is, Deutsch willingly embraces what he calls a 'no-theory theory of intuition' in which one does not need to precisely articulate what intuition is, instead one takes instances of what philosophers call intuitions and assesses their evidentiary role in philosophical theorizing. This strategy, he believes, "allows for fruitful discussion of the argumentative role of intuitions ... [and avoids] the potentially endless cycle of counterexample-and-theory-revision endemic to many attempts at conceptual analysis" (Deutsch 2015, pp. 29–30).

Here is Negative xphiles' evidence claim: "(EC) many philosophical arguments depend on treating intuitions about thought experiments and cases as evidence" (Deutsch 2015, p. 34). Deutsch draws an important distinction between the two ways of understanding this claim. It can be understood either in the sense of CONTENT intuitions (relying on intuition without considering it as evidence that it is true) or in the sense of STATE intuitions (relying on intuition as evidence which is true).

In support of his thesis, he points out that one may interpret EC in the sense of STATE intuitions as the following: "(EC1) many philosophical arguments treat the fact that certain contents are intuitive as evidence for those very contents" (Deutsch 2015, p. 36). On the other hand, one may interpret EC in the sense of CONTENT intuitions as the following: "(EC2) many philosophical arguments treat the contents of certain intuitions as evidence for or against other contents (e.g., the contents of more general principles.)" (Deutsch 2015, p. 36). Deutsch believes that EC1 is deniable and EC2 is true. The myth of the intuitive, then, is to claim that philosophers rely on the sense of STATE intuitions as evidence. Given the fact that negative xphi critique challenges only the sense of CONTENT intuitions, it is misleading to allegedly extend the critique by making it applicable to the entire philosophical practice of relying on intuitions in terms of STATE.

It follows that although Deutsch agrees that folks' intuitions about, for example, Kripke's Gödel-Schmidt case and Gettier's cases vary along with truth-irrelevant factors, he rightly doubts the philosophical significance of this variation: what is true is true, regardless of who shares it.[13, 14] Moreover, a philosopher may rely on CONTENT intuition without considering it as evidence. However, Deutsch also discusses some other methodological claims, such as philosophers relying only on intuition in support of their theorizing. Indeed, he argues that his analysis of Kripke's Gödel-Schmidt case and Gettier's cases shows that philosophers give arguments, inferences, or reasons in support of the intuitiveness of their claims. This is what he calls 'evidence for the evidence.'

In the same spirit, Timothy Williamson's view that commits negative xphi to 'judgment skepticism' is wrong. On Williamson's view, negative xphiles' attack philosophers' reliance on intuition as the evidence for the evidence. Since Deutsch argues, negative xphiles are wrong: neither philosophers rely on intuitions nor intuitions serve as the evidence for the evidence; therefore, Williamson is wrong.[15] For Deutsch, in general, a thought experiment is to be treated as evidence. We get to know that a certain thought experiment is genuine if it "depends on arguments, not intuitions;" as in every stage of Kripke's Gödel-Schmidt case and Gettier cases

[13] This criticism of experimental philosophy can be leveled a good deal with his other work being done against the very movement. See, for example, (Deutsch 2009), who argues, philosophy itself does not need to adopt the uniform of intuition.

[14] Deutsch (2010) argues for the same view: It is just a mistake built on the false assumption that intuition variability constitutes a threat to philosophy and its method.

[15] If I understand Deutsch's and Williamson's positions correctly, the difference between them lies in that Williamson considers intuition as a result of a thought experiment. Hence, although he rejects the idea that intuition has an evidential role to play in philosophy, he admits that thought experiment includes intuition or non-inferential judgement. Thus, when negative xphiles target intuition, they create judgment skepticism. Deutsch, on the other hand, thinks that philosophers give arguments, inferences, or reasons in support of the intuitiveness of their thought experiments. Therefore, for him, both Williamson's and negative xphiles' criticisms are beside the point. To my mind, Deutsch is right if Williamson was talking about real-life thought experiments. But since Williamson identifies the thought experiment with the counterfactual conditional, he believes that imaginary thought experiments can provide contestable facts, which can play the role of evidence philosophers can rely upon. It is true that, from this, it follows that he admits the existence of intuition; since he thinks that intuition is the result of the counterfactual. But, from this, it does not follow that he admits the evidentiary role of intuition. Actually, his problem is that admitting the evidentiary role of intuition psychologizes the idea of evidence that he does not support. In this context, Williamson's main problem with negative xphiles is not that they do create judgment skepticism by targeting intuition as evidence. Rather, his problem with them is that one, by a thought experiment that involves intuition, they understand the real-life ones, and two, they psychologize the idea of evidence by postulating that philosophers habitually rely on intuition as evidence.

(Deutsch 2015, p. 74).[16] To support this idea, he presents more thought experiments such as Jackson's Fred case, Searle's Chinese room case, and etc.; where philosophers rely on arguments, not intuitions as evidence. Accordingly, he concludes that "negative xphiles ... simply assume that philosophers treat intuitions as evidence and then proceed with their critique. The irony is that the empirical evidence concerning first-order philosophical practice reveals their own intuitional diversity evidence to be largely irrelevant to philosophy as it is actually practiced" (Deutsch 2015, p. 102).

Deutsch considers several responses to the negative xphi critique. One rejoinder is called the expertise reply, according to which xphiles' results are not reliable because their data consist of ordinary people's intuitions and not of philosophers' intuitions. Deutsch agrees that philosophers' judgments on philosophical matters are evidentially better than non-philosophers' ones. That is so not because philosophers are better than non-philosophers at intuiting but "because philosophers are better than non-philosophers at arguing for, and defending, philosophical judgments" (Deutsch 2015, p. 141).

Another rejoinder to xphiles is called the multiple concepts response which suggests that variation in intuition is apparent and not genuine because disagreement among people's intuitions is due to the fact that different subjects are getting different senses from the same concept. Let us say a group of subjects were asked whether or not Smith knows. Different subjects may get different senses from the concept of 'knowledge.' In reply, Deutsch states that it is the proponents of multiple concepts upon whom the burden of proof rests. Yet, both the critique and the two main replies to it are unnecessary simply because, for him, they "are predicated on a mistake, the mistake of taking it to be part and parcel of philosophy to rely on intuitions as evidence" (Deutsch 2015, p. 156). That is to say that since philosophers do not treat intuition as evidence, both the critique and the two main replies to it cut no ice in the context of philosophical argumentation.

Having presented Deutsch's main arguments, I would like to briefly comment on them as follows. Regardless of whether or not what Deutsch calls no-theory theory is promising, it indicates that he advocates an agnostic approach regarding the nature of intuition. Nevertheless, his distinction between a state intuition and a content intuition is just a theory of the nature of intuition on which he builds his whole criticism of negative xphi. One might legitimately ask the following. If, according to his no-theory theory, it is not necessary to develop a theory of the nature of intuition in order to understand its role in philosophical argumentation, what is, epistemologically speaking, the philosophical significance of the above-mentioned distinction? Either his agnostic approach is true; thus, he should not have made the distinction. Or the distinction is significant; thus, adopting a theory of the nature of intuition is necessary in order to understand its role in philosophical argumentation.

---

[16] Note that 'not' here means that there is no possibility for intuition to be considered as evidence in philosophy: Argument is the only evidence philosophers rely on. For more information about Deutsch's reasoning along this line, see (Deutsch 2016).

Either way, Deutsch is wrong. It seems that we have to think about the role of intuitions more seriously.

As was noted earlier, there is a diverse range of projects in the just considered three accounts. As a result, I will consider three major intuition-rejecting arguments from them: the argument from lacking reference, the argument from depsychologizing the philosophical evidence, and the argument from argumentationalism.

4.3 The Three Major Intuition-Rejecting Arguments

In the previous section, I looked at three main theses of anti-centrality. That is, I looked at their general accounts of rejecting the thesis that philosophers rely on intuition as evidence. It seems to me, however, that it is time now to single out the main arguments that cross-cut those accounts. In this section, I shall identify three kinds of key anti-centrality arguments that are worth considering separately: the argument from lacking reference, the argument from depsychologizing the philosophical evidence and the argument from argumentationalism.

4.3.1 The Argument from Lacking Reference

Intuition-deniers proceed with the plan of rejecting the claim that philosophers commonly rely on intuition as evidence from the starting point that evidence must be shared. Intuition varies, and accordingly, it is controversial. Therefore, intuition cannot be considered evidence.

For example, Cappelen argues that uses of intuition-terminology and cognate terms, in ordinary English as well as in contemporary philosophy, vary in an indiscriminate way. In colloquial English, intuition-talk applies either to effortless performances, hedges, sixth sense, seeming, belief, thinking, or just BLAH. Similarly, philosophers have never shown a pattern of agreement on what intuition is. Either the term was not used to constructively denote a theoretical or technical concept different from the way it was used in ordinary English. Or philosophers have given various uses of the term, such as a sui generis mental state, source of justification, belief about the abstract matter, intellectual seeming, intellectual perception, phenomenal experience, judgement, unreflective reasoning, and so on and so forth. Given the lack of consensus on what intuition is, intuition-vocabulary has no key connotation. Therefore, intuition has no significant epistemological status within philosophy: we ought not to treat intuitions as evidence.

Similarly, on Williamson's view, disagreement about the nature of intuition is a methodological scandal in contemporary analytic philosophy. Recall that he articulates this claim as follows:

> "Intuition" plays a major role in contemporary analytic philosophy's self-understanding. Yet there is no agreed or even the hoped-for correlation between our having an intuition that P and its being true that P. Since analytic philosophy prides itself on its rigor, this blank space in its foundations looks like a methodological scandal (Williamson 2007a, p. 215).

In light of that, Williamson explicitly asserts that philosophical evidence does not consist of intuitions.

In the same vein, Deutsch argues that intuition is a concept that is difficult to pin down. That is why there is a huge disagreement amongst philosophers on the nature of intuition. Any attempt to identify the concept will be responded to with many counterexamples. As a way out of such an endless cycle of counterexamples, he believes that it is more fruitful to adopt a no-theory theory of intuition.

4.3.2 The Argument from Depsychologizing the Philosophical Evidence

What is meant by psychologizing the evidence is that those who argue that philosophers rely on intuition as evidence also construct it as a folk psychological and mental state. By rejecting this psychologism, intuition-deniers aim to depsychologize the very evidence; they reject the claim of intuition as evidence and thus say that the folk psychological theory and its ascriptions of intuition as a mental state are not true.

According to Cappelen, proponents of centrality understand by intuition a psychological state. He writes: "many philosophers, including proponents of centrality, take 'intuition' to denote a psychological (mental) state or event" (Cappelen 2012, p. 8). Examples of those philosophers are "often found in the psychological literature on intuitions." They take intuitions to be "beliefs (or inclinations to believe) that are generated in a certain kind of 'spontaneous' or 'unreflective' way" (Cappelen 2012, p. 10). But Cappelen notes that his case studies offer "no evidence that judgments about cases rely on appeals to the writers' psychological states" (Cappelen 2012, 204).

Identifying intuition with a psychological phenomenon such as belief, inclination to belief, and so on prevents it, according to Williamson, from being one's epistemic evidence. He says very clearly that "our evidence in philosophy consists of facts, most of them non-psychological, to which we have appropriate epistemic access" (Williamson 2007a, p. 241). Intuition, then, does not qualify as evidence for philosophers to rest on.

Interestingly, Deutsch, in turn, thinks that what those who usually adopt a theory of the nature of intuition do is attempt to reveal specific psychological features of the relevant intuition, such as being spontaneous, non-inferential, and so on. Deutsch says that:

> …by wide consensus, being an intuition, in the sense relevant to philosophy, has something to do with the psychological features of the relevant judgments. That is, something is an intuition just in case it has certain psychological features, and one cannot say what intuitions are, in the sense of revealing their nature, without specifying these psychological

features. For example, it is sometimes said that intuitions are spontaneous or non-inferential (Deutsch 2015, p. 26).

The problem with adopting a theory of the nature of intuition, Deutsch argues, is that trying to identify intuition with any of the aforesaid psychological features does not reveal the full nature of it simply because intuition may mean more. If one identifies intuition with an inclination to belief, a cluster of counterexamples arises, identifying it with something else such as intellectual seeming, belief, etc. Keeping in mind that the way one's identifies intuition affects one's assessment of its epistemic status, Deutsch prefers to embrace the no-theory theory of intuition in which one takes instances of what philosophers commonly call intuitions and assesses their evidentiary role in philosophical theorizing.

4.3.3 The Argument from Argumentationalism

As we have seen earlier, after offering alternative interpretations of the concept of 'thought experiment,' Cappelen, Williamson, and Deutsch started underlining the role of argument and rejecting the role of intuition as evidence in philosophy.

Cappelen, for example, studies many paradigmatic thought experiments arguing that their authors argued and did not rely on intuition in them. Take, for instance, Lehrer's Truetemp case. Cappelen's claim is that, after introducing the thought experiment, Lehrer provides arguments in order to support the judgment (or what usually is considered intuition) that there is good reason to say that Mr. Truetemp does not know the temperature.

Also, Williamson assumes that philosophers call for intuitions when they do not have arguments anymore. Here is how he states the idea: "when contemporary analytic philosophers run out of arguments, they appeal to intuitions" (Williamson 2007a, p. 214). By 'argument,' as was noted earlier, he understands the counterfactual conditional in which the contestable facts play the role of evidence, as in his interpretation of Gettier's cases. In brief, philosophers rely on argument as evidence and not on intuition.

For Deutsch, in general, a thought experiment is evidence that "depends on [or consists of] arguments, not intuitions;" as in Kripke's Gödel-Schmidt case and Gettier cases (Deutsch 2015, p. 74). To support this idea, he presents more thought experiments such as Jackson's Fred case, Searle's Chinese room case, etc., and argues that their authors rely on arguments as evidence in order to support the intuitiveness in them and not the other way around.

4.4 Replying to the Three Major Intuition-Rejecting Arguments

In the previous section, I laid out the three major intuition-rejecting arguments intuition-deniers raise. In this section, I will be responding to these arguments showing how my account is not endangered by them. To be precise, those approaches just articulated do not threaten the substance of my own view of the nature of intuitions and their epistemic status defended in this book. Let us begin with responding to the first argument.

4.4.1 Reply to the Argument from Lacking Reference

As we have seen, Cappelen begins with the observation that there is a huge difference between ordinary speakers', intuition-theorists', and first-order philosophers' usage of the term 'intuition' and its cognate words. He first draws a distinction between ordinary speakers' use of intuition-terminology and intuition-theorists' usage of it. According to him, ordinary speakers, unlike intuition-theorists, almost never use intuition-terminology to refer to a mental state with at least one of the following three features: phenomenal properties, the conceptual competence of the subject, and some fundamental rock-bottom justificatory status. Indeed, ordinary speakers' usage of intuition-terminology expresses either a hedge attitude or an adjective that is fixed by the context.

After arguing for these differences between intuition-theorists' and ordinary speakers' usage of intuition-terminology, Cappelen proceeds to argue that first-order philosophers' usage of intuition-talk is more like ordinary speakers' use than intuition-theorists' usage. Most of the first-order philosophers normally do not have any of intuition-theorists' particular theoretical model of intuition in mind while working on first-order philosophical topics. He then infers that it is thus most likely that first-order philosophers' use of intuition-terminology is radically analogous to the way ordinary English use the very terminology. That is to say that both ordinary speakers' and first-order philosophers' application of the term intuition and its cognates linguistically vary according to the intuiters hedge attitude or to the context by which the adjective of the noun "intuition" is fixed.

On the other hand, though intuition-theorists themselves have little agreement on a clear definition of intuition, what makes their style of using intuition-terminology different from the other two usages is the hypothesis of intuition-theorists that intuition always comes with a special phenomenology, a conceptual competence of the subject, and fundamental rock-bottom justificatory status.

In the second chapter, I argued successfully, I believe, for the non-phenomenal character of intuition. I have shown that there is nothing it is like to be (i.e., phenomenal) intuition. And such intuition transcends any particular conceptual competence of the subject simply because one may intuit what one does not understand, and one may understand what one does not intuit. In the third chapter, I have shown that intuition alone has no justificatory status. Instead, I have confirmed that intuition is an integral part of a coherent combination associated with an epistemic status as such. With this in mind, ordinary speakers, intuition-theorists and first-order philosophers use intuition-terminology analogously.

Still, Cappelen may reply that if intuition-theorists and first-order philosophers' use of intuition-terminology is analogous to ordinary speakers, then the usage of intuition-terminology expresses a hedge attitude, an adjective that is fixed by the context, and activities with some kind of ease effortlessness, or spontaneity involved. That is, yet again intuition varies in an indiscriminate way. For example, many ordinary speakers, like several first-order philosophers, who usually might not have any model of intuition in mind while working on first-order philosophical topics, do use intuition-talk to mean a hedged attitude.

Others, like some intuition-theorists, might use it to mean spontaneous judgments. I have to admit that solving the problem of the three features helps to stop attributing the three features to intuition. But doing so does not help to stop the linguistic variation in intuition among the individuals of the above-mentioned three parties. More generally, one might argue that the argument from linguistic practice (intuition-talk) does not undermine the importance of the centrality thesis. That is, doubts about or disagreement on the concept of intuition, according to which centrality does not fit in well with philosophers' use of intuition-terminology in their texts; need not undermine the importance of intuition's status of being central to philosophy. After all, philosophers simply can rearrange their claims in terms of some other less controversial category, like 'system 1' reasoning, for example, so it is as clear and compelling as possible (Nado 2016).[17] If so, the Argument from Lacking Reference thus loses much of its force.

However, a proponent of anti-centrality like Cappelen, for instance, may object that the response just defended will not succeed in avoiding the danger from the argument from linguistic practice since, in Williamson's words, there is no agreed account of intuition. That is why, Deutsch argues, any attempt to identify the term intuition will be responded to with many counterexamples. One way to respond to this objection, however, is to admit that there is no question that philosophers disagree. Any introduction to philosophy student can attest to this fact of disagreement. Intuition is not an exception to this general statement. In light of that, creative discussions of contemporary philosophers on the nature of intuition have never shown patterns of convergence. For instance, in order to apprehend some of the more distinctive attributes of intuition, Fedyk develops a straightforward definition of intuition according to which:

[17] System 1 and System 2 are two expressions used by Daniel Kahneman (2011) to uncover the dual systems of our mind by drawing a dichotomy between two different types of thought. On Kahneman's scheme, System 1 denotes what is characterized as a fast, automatic, intuitive and mostly unconscious mode of reasoning. System 2, by contrast, denotes our slow, deliberate, analytical and intentionally effortful kind of thinking.

> an intuition is about the salient features of a case, it has propositional content, and the propositional content of an intuition is obtained in some way from the implicated concept (Fedyk 2009, p. 56).

Other philosophers consider intuition to be equivalent to a sort of belief (Lewis 1983). However, Peter van Inwagen thinks that:

> our "intuitions" are simply our beliefs-or perhaps, in some cases, the tendencies that make certain beliefs attractive to us, that "move" us in the direction of accepting certain propositions without taking us all the way to acceptance (Inwagen 1997, p. 309).

Some hold, moreover, that intuition is a unique propositional attitude in which a proposition seems true (Huemer 2001, 2005b).

In the same way, disagreement among philosophers arises when they deal with the issue of the source of intuitions. Following an established tradition in western philosophy to look for psychological origins of intuitions, Helen De Cruz states that:

> philosophical intuitions originate from spontaneous, early-developing, cognitive processes that also play a role in other cognitive domains. Additionally, they have a skilled, practiced, component (Cruz 2015, p. 233).

De Cruz thinks that intuition is best explained by the dual processing approach of our mind that draws a dichotomy between two different types of thought. On De Cruz's scheme, Type 1 denotes what is characterized as fast, automatic, intuitive and mostly implicit sort of reasoning. Type 2, by contrast, denotes our slow, explicit, deliberate, analytical and intentionally effortful kind of thinking. Intuition, then, is a production of a Type 1 mental process, and argument is a production of Type 2 mind operation.

But does the simple fact that philosophers disagree entail anything about the lacking of reference to the relevant term? Intuition, intuition-deniers think, lacks reference. Not so, if what I am going to argue for is correct. In addition to what I have argued for so far, I argue further along the line of some philosophers who have distinguished between "thin" and "thick" types of intuition. For instance, (Alexander and Weinberg 2014) write:

> But, what might such 'intuitions' be, such that they could legitimately serve these purposes? Answers vary, ranging from 'thin' conceptions that identify intuitions as merely instances of some fairly generic and epistemologically uncontroversial category of mental states or episodes to 'thick' conceptions that add to this thin base certain semantic, phenomenological, etiological, or methodological conditions (Alexander and Weinberg 2014, p. 189).

Assuming something like this distinction, one can respond to intuition-deniers by claiming that their argument is too theory-laden: they assume an excessively thick theory of intuition. The three intuition characteristics they observe, viz., intuition has a distinctive phenomenology, being based only on conceptual competence, and offering fundamental justification, are true of some thick kinds of intuition.

Contrary to that, in the previous chapters, I argued successfully, I believe, for a non-phenomenal concept of intuition that transcends any particular conceptual competence of the subject. In addition, I have shown that intuition has a justificatory status if and only if it is considered to be part and parcel of an overall entity. My concept of intuition then falls under the thin category of intuition and not the thick one. With this in mind, the myth of the intuition-deniers then is to claim that their criticisms are applicable to the thin notion of intuition. But that is not accurate.

None of the ten-thought experiment-generated intuitions considered in this book requires intuition to have any of the above-mentioned three intuition characteristics. That being so, the view of intuition defended in this book is "thin" in two senses.

First, it does not count on theoretically substantial concepts like unique phenomenology, conceptual competence, and fundamental justification. As held by my thin view, intuition has no special phenomenology. It also rejects the claim that intuition justifies one in believing a certain proposition by virtue of one's full understanding of it depending on its content and way of a combination of its concepts. Intuition as a conclusion or a premise of an epistemic argument needs to have epistemic reason(s) to produce justified belief. Thus, it is inferential, and since it itself can be such an epistemic reason, then it can also be a justifier. It need not be fundamental: justifies without being inferentially justified.

Second, it uses intuition to refer to a more general mental kind than the thick views. The object referred to by intuition, according to my thin view, is an intentional but non-phenomenal, non-conceptual, and non-propositional mental state which is individuated by its origin. And such intuition alone has no justificatory power, and it transcends the traditional a priori-a posteriori distinction.

Hence, I can agree with Cappelen, after the modification I made to his thesis, that ordinary speakers, as well as most first-order philosophers and intuition-theorists, use intuition in the same way, yet I maintain that this use of intuition does not contradict the thin picture of centrality. With a thin account of intuition, centrality continues to exist as a plausible thesis; simply because it is not the kind of intuition anti-centrality thesis was invented to deny.

On the other hand, Williamson notes that philosophers vary in identifying intuition: some of them label intuition as a priori, while others look at it as straightforwardly empirical. And some think that intuition needs to have special phenomenology. Furthermore, some think that it needs not even be non-inferential. Accordingly, he concludes that philosophers better not use the term intuition and its cognates. On my thesis, intuition is multi-dimensional: it apparently applies to any type of judgment. As we have seen, I raised doubts about the contention that intuition is either a priori or a posteriori. I argued, instead, that it may be a priori, a posteriori, both a priori and a posteriori, and neither a priori nor a posteriori. As a result, intuition cannot be explained by the traditional a priori-a posteriori dichotomy which, accordingly, has to be transcended. Intuition is better to be taken as involving several dimensions or aspects. And such intuition is non-phenomenal. In addition, I have demonstrated that intuition appears to be a crucial part of the relevant thought experiment. And thought experiment offers an epistemic argument. And for such an epistemic argument to be epistemically justified, it needs to produce justified belief both in its conclusion and premises as well. That is, intuition, whether it is a conclusion or premise, needs to have epistemic reason(s) to produce justified belief. So conceived, it is both justifier and inferential.

Max Deutsch states that philosophers largely disagree on the nature of intuition. I have admitted that philosophers disagree. But this simple fact does not entail that what philosophers do not agree on cannot be relied on as evidence. Here is one way to see the point. Plato spent a good time endeavoring to further uncover the nature of Forms, especially the form of KNOWLEDGE. The majority of contemporary philosophers, especially Gettier and his contemporaries, disagree that justified true belief is enough to be counted as knowledge. Should we, therefore, conclude that this disagreement is evidence that there is no evidence that there is something called knowledge? Certainly not. Epistemologists were perhaps mistaken about what they were doing in their attempts to discover the essence of KNOWLEGE if any, but that is not enough reason to deny that there is no evidence for any concept of knowledge. Nor is that enough reason to deny that Gettierean or *Platonistic* views can be centrally relied on as evidence. In other words, although intuition-theorists must re-describe their target within the scope of epistemic argument in a sense defined above, still their approach that philosophers do rely on intuition as evidence makes sense.

One more example from a different field, which is worth discussing in this context, is the one which shows that disagreement is not enough as a standard to measure the status of the thing disagreed upon. Suppose two directors produced two movies: one is pornographic, and the other is a documentary engaging with a philosophical movement during the Second World War. Both of the directors posted the movies on a social media platform. Suppose the pornographic movie got 93% of the views and the same percentage of the viewers' positive comments. Does this rate reflect the value of the pornographic artwork? Certainly not. The documentary artwork might be much more important, valuable, or informative than the pornographic one. In the same way, disagreement as an enumerative inductive argument may fail to reflect the epistemic status of intuition. The present example is not set to explain variability in intuitions but to show that such variability is not enough reason not to consider intuition as evidence that can be relied on. At any rate, my thesis did not consider intuition alone to have evidentiary status.

Deutsch differentiates between two ways of understanding the negative xphiles' evidence claim. Deutsch draws a distinction between CONTENT and STATE intuitions. He states that it is a myth to claim that philosophers rely on the sense of STATE intuitions as evidence. Again, there are, I think, two moves to be made here.

One is that it is true that, on the picture given in this book, the focus is on the sense of STATE intuitions, but in order for a state-like intuition to be considered as evidence, one must have such intuition, and additional conditions must be added. These additional conditions consist in that intuition is an integral part of completing a whole that cannot function without it. That being so, it is misleading to allegedly extend the anti-centrality thesis by making it apply to my particular sense of intuition as a state. To illustrate, Deutsch's argument here sounds as if he thinks that either (EC) is understood in the sense of CONTENT intuitions or the whole discipline of philosophy is undermined. However, this sort of radical claim is not able to dismiss the discipline as a whole. (EC) can be understood that experimentalists believe that philosophers rely on the sense of STATE intuitions as true evidence.

Surely, this interpretation of (EC) can have important methodological implications without being able to dismiss the discipline as a whole. Epistemologically speaking, the main target of Deutsch's criticism is xphi because it claims that its data targets philosophers' reliance on the sense of STATE intuitions as true evidence. Experimentalists are wrong because they claim that their criticisms are applicable to a thick notion of intuition. That is, the sense of STATE intuitions with a distinctive phenomenology, being based only on conceptual competence and offering fundamental justification.

Another is the supposed truth of the argument-only view of philosophical evidence, which regards intuition as a class inferior to that of argument. Those two classes cannot coexist with each other. Thus, they reject the class of intuitions wholesale. By contrast, in the picture given in this book, the focus is not only on intuition as STATE but also on intuition as epistemologists' particular sort of practice—their appeal to intuition within a whole. Doing so allows us to criticize the epistemic method (which is the argument or thought experiment according to my thesis) while the use of intuition in it as a justified belief maker remains perfectly acceptable.

4.4.2 Reply to the Argument from Depsychologizing the Philosophical Evidence

Let us turn to our second argument, which I have called the Argument from Depsychologizing the Philosophical Evidence. To start with, recall that centrality was characterized as follows: philosophers centrally rely on intuitions as evidence. This supposes that supporters of anti-centrality are concerned with the nature of evidence. And, apparently, they postulate a propositional notion of evidence. (Williamson 2002), for example, thinks that only proposition can be considered as evidence. A very similar version of understanding evidence is found in (Cappelen 2012, p. 13), who, in turn, reflects on centrality as proposing the proposition that "it is *A has the intuition* that p that serves as evidence."[18]

Deutsch (2009, 2010) makes similar points interpreting the centrality thesis along this line and claiming that it utilizes intuition-facts as premises (i.e., propositions). On Deutsch's interpretation, the term intuition was not mentioned in, for example, Gettier's thought experiment. Thus, there is no reason to assume that Gettier's thought experiment relies on an appeal to intuition as evidence.

Consider, for example, the following as a reconstruction of Frank Jackson's thought experiment of the 'Fred case' against 'physicalism.' It is intuitive that Fred's (the protagonist of Jackson's thought experiment of 'Fred case') mental states could

[18] It is noteworthy that it is unclear whether Cappelen has formulated the thesis of centrality in a way which assumes a propositional theory of philosophical evidence because, in his book (2012, p. 14), he states that he adopts a neutral position on whether only propositions can be considered as evidence in philosophy.

not be understood merely by means of physical terms. Fred's color experience shows that physicalism leaves out at least some mental states like 'qualia' or 'raw feels', for example, because our having all physical information about him won't help us to know everything about his color experience. Physicalism holds that all mental states can be understood merely by means of physical terms. Therefore, physicalism is false.

On Deutsch's reading, this reconstruction badly misunderstands the real structure of such a thought experiment. There is no reason to believe or to be inclined to believe that Jackson's thought experiment relies on such an appeal to intuition as evidence, especially since the term intuition was not mentioned in Jackson's 'Fred case.' Jackson, therefore, does not use intuition as evidence for his conclusion that physicalism is an unattractive view. Psychological facts, e.g., beliefs, inclination to believe, etc., about intuition are completely irrelevant to Jackson's conclusion. This is what I refer to as the Argument from Depsychologizing the Philosophical Evidence. In this line of thinking, (Williamson 2007a, p. 235) proposes a similar view, criticizing that philosophers have the inclination to take it that our evidence "consists of the psychological facts to the effect that we have intuitions with those contents." In the same way, Cappelen holds that there is "no evidence that judgments about cases rely on appeals to the writers' psychological states" (Cappelen 2012, p. 204). That is to say that evidence in philosophy, according to supporters of the anti-centrality thesis, consists of non-psychological propositions. More generally, therefore, Williamson's, Deutsch's, and Cappelen's positions on psychologization have skeptical worries about the connection between psychological states and propositions about the world: How could what is psychological provide evidence for what is philosophical? How have so many orthodox philosophers been misled into psychologizing the philosophical evidence?

Before proceeding further, let me make one point which is worth addressing in this context. Despite the fact that Williamson's, Deutsch's, and Cappelen's positions on psychologization resemble each other, yet, they are not identical, and a close investigation may uncover some essential differences. The main difference is that while Williamson thinks that philosophers mistakenly treat philosophical evidence as made up of intuition, Cappelen and Deutsch accept that psychologization is wrong but reject that philosophers are perpetrators of it. Williamson's position aims at revealing the nature of philosophical evidence; Deutsch's and Cappelen's positions aim at showing what philosophers believe about that very evidence.

In the previous chapters, I developed a broader view, according to which intuitions are attitudes but not propositional in nature. But suppose I am right: intuition is not propositional. So, what, why does all this matter at all? There is more than one reason. Most obviously, whether intuition is propositional or non-propositional is relevant to the question about the nature of intuition which inherits interest from the question about the epistemic status of intuition since how we answer the former affects how we answer the latter.

The account I have developed of the epistemic status of the mental state of intuition sees that the evidential status of intuition does not depend on it (i.e., intuition) itself; rather, it rests on its being combined with other pieces to make up a whole —

thought experiment or argument. The main failure of the pros and cons accounts of the evidential status of intuition is that they evaluate intuition separately. Alternatively, on my view, intuition has evidential status if and only if it plays a role in making up an epistemically justified argument. Hence, even if I might not believe that intuition is ultimate evidence, I totally agree that intuitions can fully be derivative evidence. And the postulation of centrality does not require that the evidential role of intuition ought to be ultimate or foundational. Such a version of centrality is not one which supporters of the anti-centrality plan to reject. If my non-propositional theory is correct, then intuitions can play an evidential role not by themselves but by being combined with other pieces to make up a whole.

This non-propositional view of intuition also rules out the forms of centrality, such as the "content" reading: "many philosophical arguments treat the contents of certain intuitions as evidence" (Deutsch 2015, p. 36). Another related dispute among theorists of evidence concerns the question of what evidence possession is. For proponents of propositional models of evidence, S has the evidence e if and only if S believes the truth of e. (Williamson 2002) argues that this is not enough for a subject to have evidence. To that end, he develops a stricter theory of evidence according to which in order for *e* to be a subject's evidence, *e* has to be part of her total knowledge. By doing so, Williamson "equates S's evidence with S's knowledge" (Williamson 2002, p. 185). Now, if the evidence justifies belief, then Williamson's view of the evidence = knowledge theory leads to the conclusion that "knowledge, and only knowledge, justifies belief" (Williamson 2002, p. 185). Alternatively, on my non-propositional view, in order for mental states like intuition to be a subject's evidence, she must have such intuition, and additional conditions (e.g., being combined with other pieces to make up a whole) must be added. Centrality, as understood by proponents of anti-centrality, that is, intuition plays the role of philosophers' central evidence, seems to assume that philosophers possess intuitions as genuine evidence. Not so, if my non-propositional view is correct.

While anti-Centralists are primarily concerned with contemporary analytic philosophers' usages of the intuition-terminology, it seems, however, obvious to Williamson, Deutsch, and Cappelen that philosophers treat intuition as a psychological state. A bulk of the substantial and growing meta-philosophical literature fairly endorses the claim as such. For example, Richard B. Miller claims that "intuition is the reader's own introspective judgment about his or her own linguistic disposition as manifested in the imagined circumstances" (Miller 2000, 235). As it stands, the quote implies that intuition is an introspective judgment which is a psychological state. In the same vein, (Earlenbaugh and Molyneux 2009, p. 105) suggest a different proposal "that finding P intuitive amounts to having some inclination to believe that P." That is to say that intuition, after all, is a psychological fact, namely, the inclination to believe. But Williamson's, Deutsch's, and Cappelen's reading of centrality go further, claiming that contemporary analytic philosophers' usage of intuitions is not just as psychological states but genuine psychological states. Particularly, they think that intuition is recently treated as having an evidential role or being the basic evidential source.

The necessary condition for the reliable indicators requirement of a psychological state to be categorized as basic evidential sources, according to (Goldman and Pust 1998), is: "mental states of type M constitute a basic evidential source only if M-states are reliable indicators of the truth of their contents (or the truth of closely related contents), at least when the M-states occur in M-favorable circumstances" (p. 180). What this basically means is that when a psychological state occurs in the circumstances favorable for it to occur, its contents are generally true (Goldman and Pust 1998, p. 180). Intuition is not a mere psychological state but a psychological state with an evidential role. The suggestion here is as follows. If Gettier intuits that Smith does not know, then, given the reliable connection between intuition and the truth of the proposition intuited, Gettier's intuition that Smith does not know is enough evidence for the content of the proposition 'Smith does not know,' to be true. This is not to say that the belief that Gettier intuits that Smith does not know evidentially supports the belief that Smith does not know. Rather, Gettier's intuiting that Smith does not know itself evidentially supports the belief that Smith does not know.

It is not obvious to me that intuition is treated as a genuine psychological state. Throughout the literature on intuition, there have been various attempts to characterize intuition as something different from a psychological state. For example, (Hogarth 2001) has a tendency to characterize intuition as similar to perception. We deal with intuition as seeing the truth of some proposition, just as we deal with perception as seeing some state of affairs. And (Richard 1990) describes intuition as a propositional attitude as defined earlier by the that-clause. After all, there are many other vivid examples of conceptual work to define the nature of intuitions as several other things rather than being a psychological state. From these considerations, I maintain that intuition is not always characterized as a genuine psychological state.

There are two points to be made here. First, we just have seen that there is no agreement to identify intuition as a psychological state. If anti-centralists want to deny this disagreement, it entails that their general standard for something to be considered as evidence, it must be commonly shared, is wrong. This suggests that their Argument from Depsychologizing the Philosophical Evidence is untrue. In particular, if Williamson's denial of the principle of "Evidence Neutrality" is true, then the Argument from Depsychologizing the Philosophical Evidence is untrue. According to Williamson, philosophical evidence does not consist of the principle of "Evidence Neutrality," which he defines as follows:

> Whether a proposition constitutes evidence is in principle uncontentiously decidable, in the sense that a community of inquirers can always in principle achieve common knowledge as to whether any given proposition constitutes evidence for the inquiry... in a debate over a hypothesis h, proponents and opponents of h should be able to agree whether some claim p constitutes evidence without first having to settle their differences over h itself. (Williamson 2007a, p. 210).

What this quotation means is that evidence is always known, which in turn means that one can always acquire knowledge if one overcomes all the "accidental mistakes and confusions" (Williamson 2007a, p. 210). This is to say that disagreement about intuition is motivated by accidental mistakes and confusion that can be overcome. Once they are overcome, there will be agreement about intuition, and accordingly, it will be considered evidence. If so, it is better that people should not believe in the Argument from Lacking Reference. And, if the disagreement cannot be overcome, then, as we have seen earlier, there is no agreement to define intuition as a psychological state. In this case, it is certainly better that people should not believe in the Argument from Depsychologizing the Philosophical Evidence. So, either way, it is better that people should not believe in the anti-centrality thesis.

Second, on the picture given in this book, intuition-based epistemology can avoid the charge from psychology. On the suggested view, intuition is a mental state of an epistemological kind rather than a faculty that picks out special psychological mechanisms, such as being phenomenal, propositional, etc. Recall that I have argued that intuition cannot be individualized in terms of being propositional, phenomenal or in terms of special feeling or what it feels like, emotional, moody, and dispositional mechanisms. Having rejected all these psychological mechanisms, does this mean that the mental state of intuition is counted as one's evidence only if she has this mental state? No. At least, this is not what I have argued for. Again, merely having the mental state of intuition is not enough to possess it as evidence. And therefore, additional conditions need to be added. The suggestion is that intuition is an integral part of completing a whole. And it is individuated in terms of its origin, namely, the thought experiment, which is considered to be an intuition pump.

Being appealed to by thought experiments, both intuitions and thought experiments are supposed to have a lot in common. In light of this fact, I have developed an account of the epistemic status of intuitions. By witnessing the failures of the arguments of reliablists, skeptics, and perspectival relativism regarding the epistemic status of intuitions, I see that the evidential status of intuition does not depend on it itself, rather it rests on its being combined with other pieces to make up a whole — thought experiment or argument. The main failure of the aforesaid pros and cons accounts of the evidential status of intuition is that they evaluate intuition separately. Alternatively, on my view, intuition has evidential status if and only if it plays a role in making up an epistemically justified argument. There is thus no need to retreat into psychological facts to advocate the epistemic status of intuitions. I then claim that, though the Argument from Depsychologizing the Philosophical Evidence might work against other accounts of intuition, my account remains untouched.

In sum, my approach does not rest on intuition being distinctive of philosophical methodology. The only crucial fact is that it rests on the epistemic thought experiment, which, in turn, is structured as the epistemic argument from which epistemic intuition arises. If so, the question of epistemic status does not apply to intuition separately. Rather, it is applicable to the argument itself in which intuition has evidential status if and only if it plays a role in making up the epistemically justified argument. This is one way, then, where arguments against centrality do show that there is further work to be done in articulating the anti-centrality strategy. But surely, the approach as articulated now does not threaten my own view of intuition.

4.4.3 Reply to the Argument from Argumentationalism

Another line of attack on centrality holds the claim that philosophers use argument rather than intuition in their evaluation of philosophical theories and hypotheses. As we have seen earlier, each of Cappelen, Williamson, and Deutsch separately has defended a similar approach in order to refute centrality. To remind the reader, they analyzed a series of canonical epistemic thought experiments such as Gettier cases, the Truetemp case, Kripke's Gödel–Schmidt case and so on in the light of this approach. As a result, they concluded that this is the right conception to adequately understand thought experiments in philosophy. In addition, they held that centrality is a misconstruction of philosophers' practice and that intuition rarely, if ever, plays a sufficiently great evidential role in philosophy. Call the above argument against centrality the argument from argumentationalism. In this section, I will mainly focus on three canonical epistemic thought experiments analyzed by intuition-deniers where I find their proposed characterization to be problematic and hence does not show that my view is wrong.

4.4.3.1 Cappelen's Discussion of Chalmers' Zombie Case

Let us look back on Chalmers' Zombie case. Recall that Chalmers displays the Zombie example against physicalism. The assumption is that except for lacking conscious experience, a Zombie is identical to a normal human being. Remember that I have reconstructed the case as follows. If physicalism, the view that physical stuff is enough to explain consciousness, is right, then there should be no possibility for there to be a world that is physically identical to but lack some components (e.g., consciousness) of ours. But a Zombie, which is a creature that is physically our duplicate but lacks our conscious experiences, is conceivable. Therefore, physicalism is wrong.

Cappelen's main strategy consists of two points. One is that what most intuition-theorists consider to be three typical features of intuition are absent in the case. Another is that Chalmers argued for the conclusion of the case and accordingly relied on an argument, not intuition.

It is always a bit odd to claim that intuition and argument are mutually opposed, inconsistent or exclusive. Part of the strategy of my thesis rests on the idea that there is also the argument-friendly view of intuition, not discussed by positions on the nature and epistemic status of intuition but developed and defended in this book. Intuition is a piece of a whole —a thought experiment or argument— and helps in constructing it and shares its epistemic status. So, it came as no surprise to me that Cappelen finds arguments in the world of canonical thought experiments simply because I believe that intuition figures in these arguments. Having discussed the issue of three typical features of intuition multiple times, I will mainly focus on the second point, arguing that intuition figures in the argument.

The obvious candidate step in reasoning is that 'the world of Zombie is conceivable.' Indeed, conceivability judgments appear to be a compelling kind of justification for a belief. Cappelen's move is to separate out pre-theoretic judgment from intuition, arguing that we might have the former without certainly having the latter. The key idea at play in the Zombie case is conceivability under ideally rational reflection. He says:

> what is conceivable on ideal reflection is not something that we can settle without careful reflection and argumentation and so will be a conclusion we reach inferentially ... There's an interesting phenomenon that we pre-theoretically would characterize as the prima facie conceivability of a zombie world (Cappelen 2012, p. 186–187).

Pre-theoretic judgment is significantly important both for experimentalists' projects and philosophical argumentation. We cannot ignore that pre-theoretic conceivability is what underlies ideal conceivability. If Chalmers, for example, did not have this intuition or pre-theoretic judgment of the conceivability of a Zombie world, he would not be provided with a motive to find out whether or not this conceivability withstands careful examination under ideal conditions. What is conceivable, he writes:

> is conceivable on ideal rational reflection ...What is conceivable on ideal reflection is not something that we can settle without careful reflection and argumentation and so will be a conclusion we reach inferentially (Cappelen 2012, p. 186).

In light of this account of what is ideally conceivable, Cappelen makes the mistake of supposing that since Chalmers presents an argument to support his view, intuition must be absent. Even when one is ideally a rational conceiver, she perhaps should reflectively reason, trying only to build up epistemically justified arguments in the way explained before. Some truths will strike such an ideal agent as intuitive, playing a significant role in setting up the gradual accumulation of epistemically justified arguments as such. The suggestion is that intuition alone has no epistemic status simply because intuition appears to be a crucial part of the relevant thought experiment. The thought experiment offers an epistemic argument. Therefore, the question of epistemic status does apply to the argument buttressed intuition. This version of centrality holds that philosophers rely on an argument that includes intuition as a crucial part of it. This mutual action or influence between argument and intuition is cordial, even integral, not competitive or reciprocally excluding.

4.4.3.2 Williamson's Interpretation of Gettier's Ten Coins Case

In the face of the claim that intuition was often used as evidence in philosophy, Williamson leads another line of attack on centrality, arguing that it is ill-motivated. He involves the claim that philosophers routinely use arguments rather than intuition as evidence in their philosophizing. To illustrate this approach in refuting centrality, he offers an interpretation of Gettier's ten coins thought experiment, arguing that Gettier does not actually use intuition as evidence against the conception of knowledge as justified true belief. Remember that his analysis does not concern what Gettier's case, as a canonical example, is. Instead, his concern is how we must adequately understand such a thought experiment. In very rough terms, the Gettier case must be understood as a counterfactual conditional. Accordingly, he interprets the results of counterfactual conditional, which has been habitually expressed by what is called Gettier's intuition as follows. If a thinker were Gettier-related to a proposition, he/she would have justified true belief in it without knowledge. He goes from a modal premise (If a thinker were Gettier-related to a proposition) to a modal conclusion (then he/she would have justified true belief in it without knowledge).

So conceived, Williamson concludes that Gettier's case is possible, not that it is actual. Add to that that intuition provides access to psychological facts not to contestable facts, and thereby it fails to play the role of evidence in philosophy. Therefore, Williamson concludes that the adequate way to investigate the subject matter of philosophy is based on argument. In particular, argument or counterfactual thought experiment without intuition is what provides epistemic access to contestable facts as such.

Williamson's view, generally speaking, is true depending on whether his way of thinking about a possible world or counterfactual situation is true. One looks at "a possible world as if it were like a foreign country. One looks upon it as an observer" (Kripke 1980, p. 43).[19] One can observe whether or not Gettier has moved to the other country only in terms of qualities but not in terms of properties. One observes that someone is bald, for instance, but not someone is Gettier. However, in general, this does not seem to be the right way to look at the possible worlds. In Kripke's words: "a possible world isn't a distant country that we are coming across, or viewing through a telescope" (Kripke 1980, p. 44). To be precise, "'possible worlds' are total 'ways the world might have been' or states or histories of the entire world" (Kripke 1980, p. 18). In this sense, "the actual world is one of the possible worlds" (Kripke 1980, p. 38). Consequently, "we identify objects across possible worlds as objects resembling the given one in the most important respects" (Kripke 1980, p. 53). That is to say that an essential property designates the same object in every possible world. So perceived, we have enough reason to move towards looking upon worlds both actual and possible as they have slight differences if any. They are analogous in the most important respects.

Applied to Gettier's ten coins case, the property that 'Gettier has an intuition that Smith does not know the proposition that the one who will get the job has ten coins in his pocket is true' identifies him as 'his subject, Smith, has a justified true belief without knowledge,' in every possible world. The basic point is that there is a direct, natural implication of this resemblance, which is that Williamson's idealization of the relationship between possibility and actuality can no more bar the actuality of Gettier cases so that they can actually take place. That is to say that one is no more required to refrain from intuition simply because actuality and possibility are similar. So, possibilities cannot fall short of appropriately guaranteeing their

[19] Note that Kripke recommends using the term 'possible world' in the same sense as 'possible state (or history) of the world' or 'counterfactual situation.' What interests us here is that 'possible world' and 'counterfactual situation,' in his context, were used as synonyms.

justifiedness. As a result, some intuitions may successfully constitute justified belief, and accordingly, their thought experiment will be epistemically justified in the way defined in this work.

4.4.3.3 Deutsch's Interpretation of Kripke's Gödel-Schmidt Case

Deutsch (2009, 2010, 2015, 2016)'s investigation of the Gödel-Schmidt case has produced many important results. Contrary to those who think that Kripke relied on intuition as evidence to underpin his Gödel-Schmidt case against descriptivism, Deutsch suggests that Kripke has actually used further evidence, namely, arguments to support thought experimental judgments where "Gödel" doesn't refer to Schmidt. More specifically, he holds that Kripke presents in the original presentation of the Gödel-Schmidt the following arguments for the thought experimental judgment: "Gödel" doesn't refer to Schmidt.

The first argument is the Einstein example. Although Einstein is not the inventor of the atomic bomb, and regardless of the fact that some people associate with Einstein, the only description, i.e., the inventor of the atomic bomb, Einstein does not refer to the inventor of the atomic bomb.

The second argument is the 'immunity to error' view. The argument goes as follows. If descriptivism is right, then it entails that people can never commit errors when they emphasize the thought experimental judgment 'Gödel is the prover of incompleteness.' Since this result is wrong, there is good reason to believe that it is certainly better that people should count that descriptivism is wrong too; however, descriptivism is "the only reason to make the opposing judgment [i.e. "Gödel" refers to Schmidt]" (Deutsch 2015, p. 110). The immunity to error argument thereby grounds an indirect argument for the thought experimental judgment: "Gödel does not refer to Schmidt. For the very same reasons, Kripke's negative arguments against the descriptivist theory of reference and positive arguments for his own causal historical theory of reference also count as indirect arguments for the thought experimental judgment: "Gödel" does not refer to Schmidt.

Deutsch's main suggestion is that both of the above-mentioned arguments are not based on starting points of folks' intuitions. Their starting points concern facts about reference and not folks' intuitions about those facts. For that reason, he concludes that centrality is an incorrect view because it is based on a faulty understanding of philosophers' practice and that intuition rarely, if ever, plays significant evidential roles in philosophy. In what follows, I will show that Deutsch does not provide any good argument against the version of centrality defended in this book. Here is my strategy.

In light of the above, the following point is central to Deutsch's argument. Philosophers rarely count folks' intuitions as the starting points for their arguments. The main reason why Deutsch thinks that a claim as such is correct is the fact that philosophers very occasionally use the term "intuition" and its cognates in their original texts, as in Kripke's Gödel-Schmidt case. Apparently, he expects that a philosopher who uses intuitions as evidence must explicitly mention things like

'Look here, I am starting my argument with the intuition that so and so!' For example, he expects Kripke to explicitly mention that he is starting his argument with the intuition that "Gödel" does not refer to Schmidt.

Admittedly, sometimes one might implicitly use the starting point in one's argument. However, if philosophers sometimes start their arguments from implicit reports relating to intuitions, then it is unlikely that intuition-terminology would explicitly happen so often in philosophical practice. In this sense, we can agree with Deutsch that philosophers rarely count folks' intuitions as the starting points for their arguments. However, I fail to see why we have to trust that this is the only way in which intuition can play a central evidential role. If the implicit use of intuition in one's argument is true, then intuitions most likely are used as evidence in the arguments, even if intuition-language does not explicitly appear in those arguments. Centrality thus remains cogent despite Deutsch's argument, assuming that intuition as evidence is explicit in its nature.

Applied to the canonical thought experiments restructured as arguments in this book, it seems just obvious that in order for the fact that we have an intuition, which figures in an argument as our evidence, we do not necessarily have to explicitly appeal to it. Take, for instance, Frank Jackson in his Fred case against physicalism. With his sentence, "it seems that no amount of physical information about Fred's brain and optical system tells," Jackson did refer to the fact that he had an intuition that our having all physical information about Fred won't help us to know everything about his color experience (Jackson 1982, p. 129). This intuition is treated as a strong justification against physicalism. The same story goes for his sentence, "it seems, however, that Mary does not know all there is to know" (Jackson 1986, p. 291). Jackson did refer to the fact that he has the intuition that Mary learns something new though this intuition is considered a strong justification against physicalism.

To see this, recall the concept of epistemic thought experiment or argument defended in this book. In evaluating an epistemic argument, we need to meet the standard of justifiedness, according to which an argument is epistemically good if and only if it produces justified beliefs both in its conclusion and premises as well. For example, the sentence "it is just that none of this functioning will be accompanied by any real conscious experience" justifies the fact that Chalmers has the intuition that the realm of Zombie or Zombie itself (the subject of the story) is, indeed, conceivable (Chalmers 1996, p. 95). In a second step, this intuition is taken as justification against another distinct form of physicalism according to which consciousness can be reductively knowable in a merely physical sense.

It seems that in order for the fact that we have an intuition to figure in a justified argument as our evidence, we do not necessarily have to explicitly appeal to it. Applied to Cohen's Lottery cases, his first case's sentence, "it does not seem right to say that S knows he will lose," justifies the belief that Cohen has the intuition that the subject does not know that he will lose. And his second case's sentence, "we are inclined to say that S does know that he loses," strongly justifies believing the fact that Cohen has the intuition that the subject knows that he will lose (Cohen 1988, p. 92). Those intuitions, in the second step, are counted as justification for the conclusion that induction is not enough ground for obtaining true knowledge. In the same vein, Lehrer's sentence: "Surely not," gives a justification in believing the fact that Lehrer has the intuition that the subject, Mr. Truetemp, does not know what the temperature is when he truly says what it is (Lehrer 1990, p. 164). This intuition itself, in a second step, is considered as evidence against the externalist version of reliabilism.

The point is basically this. In epistemic arguments, sometimes one does not lingually mention that one has grounded intuition from which a premise or conclusion may be inferred or followed, but in most cases, one does hold beliefs about the hidden grounds one uses to justify what follows. To emphasize the point a bit further, consider Goldman's two sentences: "given this information, would we say that Henry knows that the object is a barn? Most of us would have little hesitation in saying this" and "given this new information, we would be strongly inclined to withdraw the claim that Henry knows the object is a barn" (Goldman 1976, pp. 772–3). These sentences are the grounds from which the intuitions that Henry knows/does not know are inferred or followed. Goldman does hold beliefs about these inferred intuitions and uses them to justify what follows: the claim that what determines whether our assessment is true or false is the distinguishability or discriminability of the situation.

By the same token, Gettier's intuition that Smith does not know the proposition the one who will get the job has ten coins in his pocket is true is given by his sentence, "it is equally clear that Smith does not know" (Gettier 1963, p. 122). After that, Gettier uses the very intuition as evidence against the justified true belief theory of knowledge. In like manner, Kripke establishes the intuition that we do not refer to Schmidt when we use the name "Gödel":

> so, since the man who discovered the incompleteness of arithmetic is in fact Schmidt, we, when we talk about 'Gödel', are in fact always referring to Schmidt. But it seems to me that we are not. We simply are not (Kripke 1980, p. 84).

Then, he considers it as evidence against descriptivism. Finally, BonJour's intuition that the subject Norman is epistemically unjustified in believing that the President is in New York City originates from the following phrases:

> is Norman epistemically justified in believing that the President is in New York City, so that his belief is an instance of knowledge? According to the modified externalist position, we must apparently say that he is. But is this the right conclusion? Aren't there still sufficient grounds for a charge of subjective irrationality to prevent Norman from being epistemically justified? (BonJour 1985, p. 41).

In the second step, such an intuition is then taken as evidence against externalism.

In virtue of the above, we can refer to intuition without using the term 'intuition.' We can also rely on intuition, being combined with other pieces to make up a whole argument, as evidence without explicitly appealing to it. The role intuitions play as evidence can neither be simply read from our words nor from the premises of our arguments in a logical sense. I, therefore, take it that my approach cannot easily be undermined and has not been undermined by Deutsch's argument.

An objection from Cappelen that needs to be considered is the following:

a number of paradigmatic cases/thought experiments from various subfields of contemporary philosophy ... the overall conclusion is that there is no 'implicit' reliance on intuitions in philosophical practice" (Cappelen 2012, p. 96).

In reply, one can say that what Cappelen actually did is he identified the above-mentioned three features of the intuitive. And then, he started looking at ten typical thought experiments to see whether there was any appeal to those features. There is, he found, no appeal to all or one of those features. Therefore, he inferred that there is no reliance on intuition, as evidenced in philosophers' thought experiments. Admittedly, he won't find any simply because intuition has none of those features, as I already argued by providing a proposal about what intuition is. Cappelen has been operating on the surface level—maybe things would look different had he investigated the real form intuition has despite the lack of those features. On the other hand, he is concerned with how to detect the presence of intuition that can operate separately. As I pointed out earlier, instead, we have to think of intuition as part of a whole or argument. Thus, the question of epistemic status does not apply to intuition separately. Rather, it is applicable to the argument itself, which can be either epistemically justified or unjustified.

4.5 Concluding Remarks

As stated by (Kornblith 2006), the greatest hope of philosophy is to uncover the principal truths about the world. Such truths can be arrived at by means of our intuitions. However, the minute we constitute the correct account intended to explain the nature of intuitions along with their epistemic status and accept that we can arrive at those noble truths, we can fulfill that ambition to which philosophy has aimed at. Realizing the nature of intuition correctly and weighing intuitions appropriately gives us the tools to avoid the pitfalls of the proponents of anti-centrality while keeping the goals.

Conclusion

The main theme of this book is to show that intuition-theorists must be careful while examining their meta-philosophical assumptions regarding both the nature and epistemic status of intuitions. Even the reliablists' widely accepted assumption that the modal character of some intuitions makes every intuition particularly logical and mathematical deserves scrutiny. I have argued that the mental state of intuition which arises from epistemic thought experiments is different from intuition as it relates to logical and mathematical objects. Unless we realize this difference, we are prone to jump to Platonism about intuition.

Reliablists have offered some arguments for the view that intuition, in the rationalist sense, is a priori. Experimentalists have claimed that, by surveying responses to cases, intuition is a posteriori. Alternatively, I have offered some convincing reasons why both of the views are misleading. In fact, we should not explain intuition in terms of either-or fallacy, either a priori or a posteriori. Instead, I have argued successfully, I believe, that intuition is multi-dimensional. I have also urged that correctly understanding the nature of intuition can be decoupled from adopting a thick notion of it. One might think that since reliablists characterize intuition as a phenomenal mental state and a propositional attitude, then this is a sufficient justification for us to admittedly characterize our intuitions in this way. However, intuition in the thin sense plausibly plays a central role in this book. I have explained in detail that we cannot refer to the mental state of intuition as having any of the following features: phenomenal properties, the conceptual competence of the subject, and a propositional attitude. This conclusion leads to an alternative, according to which intuition is individuated by the thought experiment from which it arises.

Apart from this, I have argued that the evidentiality of intuition remains plausible, but it is important to keep in mind that there are some theoretical limitations. For instance, we should not understand intuition alone as playing the central evidential role in the epistemological inquiry. Keeping in mind that I have argued that intuition is non-propositional, the view that philosophical evidence consists of intuition separately seems to be untenable. For that reason, I have thought of thought

M. Hamdo, *Epistemic Thought Experiments and Intuitions*, Philosophical Studies Series 150, https://doi.org/10.1007/978-3-031-33480-1

experiments as offering an epistemic and not a logical argument in which intuition figures and accordingly, the evaluation of such an argument is best understood in epistemological terms. So conceived, intuition plays an evidential role by virtue of its ability to fit in well with other pieces to construct a justified epistemic argument. My concept of the thin notion of intuition is plausible only if we adopt the view according to which intuition alone has no justificatory status, rather; intuition's justificatory power is derived from its possession of epistemic reason(s) to produce justified belief in the conclusion and premises of an epistemic argument.

Lastly, I have addressed standard challenges to the account of intuition that underwrites my theory of intuition, namely challenges of anti-centrality, according to which philosophers do not regard intuition as evidence in philosophy. Supporters of the anti-centrality response have offered some convincing reasons against the view that intuition, in the thick sense, does not occupy any central evidential status in epistemology. More precisely, they mainly offered three major arguments: the argument from lacking reference, the argument against psychologizing the philosophical evidence, and the argument from argumentationalism.

However, I have discussed and responded to them, showing that they fail to affect the account of intuition I attempted to establish in this book. I have supported my version of centrality by arguing against the first objection. It was shown that philosophers can always create an apt for less controversial terminology to express what is intuitive and what is not. I have also urged that my version of centrality is plausible since I adopted a thin notion of intuition which can be considered as evidence if and only if one has such intuition with additional conditions. Regarding the second objection, I have shown that it is not applicable to my version of centrality that adopts a theory of evidence which is neither propositional nor psychological in the way defined by anti-centralists. Lastly, I have argued that actual and possible worlds are slightly different, offering an argument-friendly approach to intuition. Therefore, my account is also secured from the third objection.

However, most of the attempts that one tries are open to some interesting counterexamples or other objections. This book is not an exception. So, there may be cases that question this book so as to oppose it by presenting some type of counterexample or the other. I do not know how such cases would go, but they should not, I hope, be such as both to make the thesis of this book trivial and false. I think that what I have said in this book is the right account of the nature of intuition and its role in epistemology. I hope I have done enough in the book to make others also think so.

Bibliography

Adleberg, T., M. Thompson, and E. Nahmias. 2015. Do Men and Women Have Different Philosophical Intuitions? Further Data. *Philosophical Psychology* 28 (5): 615–641.

Alexander, J., and J.M. Weinberg. 2014. The Challenge of Sticking with Intuitions Through Thick and Thin. In *Intuitions*, ed. B.A. Robert and P.R. Darrell, 187–212. New York: Oxford University Press.

Arthur, R.T.W. 2018. Thought Experiments in Newton and Leibniz. In *The Routledge Companion to Thought Experiments*, ed. M.T. Stuart, Y.J.H. Fehige, and J.R. Brown, 111–127. London: Routledge.

Audi, R. 2008. Intuition, Inference, and Rational Disagreement in Ethics. *Ethical Theory and Moral Practice* 11: 475–492.

Ballantyne, N., and E.J. Coffman. 2011. Uniqueness, Evidence, and Rationality' Philosophers. *Philosophers' Imprint* 11: 1–13.

Barsalou, L.W. 1999. Perceptual Symbol Systems. *Behavioral and Brain Sciences* 22 (4): 577–660.

———. 2003. Situated Simulation in the Human Conceptual System. *Language and Cognitive Processes* 18 (516): 513–562.

———. 2009. Simulation, Situated Conceptualization, and Prediction. *Philosophical Transactions of the Royal Society B: Biological Sciences* 364 (1521): 1281–1289.

Barsalou, L.W., W.K. Simmons, A.K. Barbey, and C.D. Wilson. 2003. Grounding Conceptual Knowledge in Modality-Specific Systems. *Trends in Cognitive Sciences* 7 (2): 84–91.

Barz, W. 2019. The Aporetic Structure of Philosophical Problems. *Journal of Didactics of Philosophy* 3 (1): 5–18.

Bealer, G. 1982. *Quality and Concept*. New York: Oxford University Press.

———. 1987a. The Philosophical Limits of Scientific Essentialism. *Philosophical Perspectives* 1: 289–365.

———. 1987b. The Boundary Between Philosophy and Cognitive Science. *Journal of Philosophy* 84: 553–555.

———. 1992. The Incoherence of Empiricism. *Proceedings of the Aristotelian Society* 66: 99–143.

———. 1993. Materialism and the Logical Structure of Intentionality. In *Objections to Physicalism*, ed. H. Robinson, 101–126. Oxford: Oxford University Press.

———. 1994. Mental Properties. *The Journal of Philosophy* 91: 185–208.

———. 1996a. A Priori Knowledge and the Scope of Philosophy. *Philosophical Studies* 81: 121–142.

———. 1996b. On the Possibility of Philosophical Knowledge. *Philosophical Perspectives, Metaphysics* 10: 1–34.

M. Hamdo, *Epistemic Thought Experiments and Intuitions*, Philosophical Studies Series 150, https://doi.org/10.1007/978-3-031-33480-1

———. 1996c. "A Priori" Knowledge: Replies to William Lycan and Ernest Sosa. *Philosophical Studies* 81: 163–174.
———. 1998a. Intuition and the Autonomy of Philosophy. In *Rethinking Intuition: The Psychology of Intuition and its Role in Philosophical Inquiry*, ed. M. DePaul and W. Ramsey, 201–240. Lanham: Rowman and Littlefield.
———. 1998b. A Theory of Concepts and Concept Possession. *Philosophical Issues* 9: 261–301.
———. 1998c. Concept Possession. *Philosophical Issues* 9: 331–338.
———. 1998d. Propositions. *Mind* 107: 1–32.
———. 1999a. A Theory of the a Priori. *Philosophical Perspectives. Epistemology* 13: 29–55.
———. 1999b. The a priori. In *The Blackwell Guide to Epistemology*, ed. J. Greco and E. Sosa, 243–270. Oxford: Blackwell.
———. 2002. Modal Epistemology and the Rationalist Renaissance. In *Conceivability and Possibility*, ed. T.S. Gendler and J. Hawthorne, 71–125. New York: Oxford University Press.
———. 2004. The Origins of Modal Error. *Dialectica* 58: 11–42.
———. 2008. Intuition and Modal Error. In *Epistemology: New Essays*, ed. Q. Smith, 189–223. Oxford: Oxford University Press.
Becker, A. 2018. Thought Experiments in Plato. In *The Routledge Companion to Thought Experiments*, ed. M.T. Stuart, Y. Fehige, and J.R. Brown, 44–56. London: Routledge.
Beebe, J.R. 2004. Reliabilism, Truetemp and New Perceptual Faculties. *Synthese* 140: 307–329.
Bengson, J. 2013. Experimental Attacks on Intuitions and Answers. *Philosophy and Phenomenological Research* 86 (3): 495–532.
———. 2014. How Philosophers use Intuition and 'Intuition'. *Philosophical Studies* 171 (3): 555–576.
———. 2015. The Intellectual Given. *Mind* 124 (495): 707–760.
Ben-Yami, H. 1997. Against Characterizing Mental States as Propositional Attitudes. *The Philosophical Quarterly* 47 (186): 84–89.
Berríos, G.E. 1989. What is phenomenology? A review. *Journal of the Royal Society of Medicine* 82 (7): 425–428.
Biro, J., and H. Siegel. 1992. Normativity, Argumentation, and an Epistemic Theory of Fallacies. In *Argumentation Illuminated*, ed. F.H. van Eemeren, R. Grootendorst, J.A. Blair, and C.A. Willard, 81–103. Amsterdam: SICSAT.
———. 1997. Epistemic Normativity, Argumentation, and Fallacies. *Argumentation* 11 (3): 277–292.
Blair, J.A., and R.H. Johnson. 1987. Argumentation as Dialectical. *Argumentation* 1: 41–56.
Bokulich, A., and M. Frappier. 2018. On the identity of thought experiments: thought experiments rethought. In *The Routledge Companion to Thought Experiments*, ed. M.T. Stuart, Y.J.H. Fehige, and J.R. Brown, 545–557. London: Routledge.
BonJour, L. 1985. *The Structure of Empirical Knowledge*. Cambridge: Harvard University Press.
Braddon-Mitchell, D., and F. Jackson. 2007. *Philosophy of Mind and Cognition*. 2nd ed. Oxford: Blackwell.
Brandt, R.B. 1979. *A Theory of the Good and the Right*. Oxford: Oxford University Press.
Brendel, E. 2004. Intuition Pumps and the Proper Use of Thought Experiments. *Dialectica* 58 (1): 89–108.
———. 2018. The Argument View: Are Thought Experiments Mere Picturesque Arguments? In *The Routledge Companion to Thought Experiments*, ed. M.T. Stuart, Y.J.H. Fehige, and J.R. Brown, 281–292. London: Routledge.
Brewer, T. 2006. Three Dogmas of Desire. In *Values and Virtues: Aristotelianism in Contemporary Ethics*, ed. T. Chappell, 253–285. Oxford: Clarendon Press.
Brown, J.R. 1986. Thought Experiments Since the Scientific Revolution. *International Studies in the Philosophy of Science* 1: 1–15.
———. 1991. *The Laboratory of the Mind: Thought Experiments in the Natural Sciences*. London: Routledge.
Brueckner, A., and A. Bundy. 2012. On "Epistemic Permissiveness". *Synthese* 188: 165–177.

Brun, G. 2018. Thought Experiments in Ethics. In *The Routledge Companion to Thought Experiments*, ed. M.T. Stuart, Y.J.H. Fehige, and J.R. Brown, 195–210. London: Routledge.
Buchanan, R. 2012. Is Belief a Propositional Attitude? *Philosophers' imprint* 12: 1–20.
Buckwalter, W., and S. Stich. 2013. Gender and Philosophical Intuition. In *Experimental Philosophy*, ed. J. Knobe and S. Nichols, 307–346. New York: Oxford University Press.
Burge, T. 1979. Individualism and the Mental. *Midwest Studies in Philosophy* 4 (1): 73–121.
———. 1982. Two Thought Experiments Reviewed. *Notre Dame Journal of Formal Logic* 23 (3): 284–294.
Buzzoni, M. 2018. Kantian Accounts of Thought Experiments. In *The Routledge Companion to Thought Experiments*, ed. M.T. Stuart, Y.J.H. Fehige, and J.R. Brown, 327–341. London: Routledge.
Cappelen, H. 2012. *Philosophy Without Intuitions*. Oxford: Oxford University Press.
———. 2014a. Précis of Philosophy Without Intuitions. *Philosophical Studies* 171 (3): 513–515.
———. 2014b. Reply to Boghossian, Brogaard and Richard. *Analytic Philosophy* 55 (4): 407–421.
Cappelen, H., and D.G. Winblad. 1999. "Reference" Externalized and the Role of Intuitions in Semantic Theory. *American Philosophical Quarterly* 36 (4): 337–350.
Casullo, A. 2013. Articulating the A Priori–A Posteriori Distinction. In *The a Priori in Philosophy*, ed. A. Casullo and T.C. Joshua, 249–273. Oxford: Oxford University Press.
Casullo, A., and J.C. Thurow. 2013. Introduction. In *The a Priori in Philosophy*, ed. A. Casullo and J.C. Thurow, 1–8. Oxford: Oxford University Press.
Chalmers, D.J. 1996. *The Conscious Mind: In Search of a Fundamental Theory*. New York: Oxford University Press.
Christensen, D. 2007. Epistemology of Disagreement: The Good News. *Philosophical Review* 116 (2): 187–217.
———. 2009. Disagreement as Evidence: The Epistemology of Controversy. *Philosophy Compass* 4 (5): 756–767.
Chudnoff, E. 2011a. What Intuitions are Like. *Philosophy and Phenomenological Research* 82: 625–654.
———. 2011b. The nature of intuitive justification. *Philosophical Studies* 153: 313–333.
———. 2011c. What Should a Theory of Knowledge Do? *Dialectica* 65 (4): 561–579.
———. 2012a. Presentational Phenomenology. In *Consciousness and Subjectivity*, ed. S. Miguens and G. Preyer, 51–72. Berlin: De Gruyter.
———. 2012b. Is Intuition Based On Understanding? *Philosophy and Phenomenological Research* 86 (1): 1–26.
———. 2013a. *Intuition*. Oxford: Oxford University Press.
———. 2013b. Intuitive Knowledge. *Philosophical Studies* 162: 359–378.
———. 2013c. Awareness of Abstract Objects. *Noûs* 47 (4): 706–726.
———. 2014a. Intuition in Mathematics. In *Rational Intuition: Philosophical Roots, Scientific Investigations*, ed. B. Held and L. Osbeck, 174–191. New York: Cambridge University Press.
———. 2014b. The Rational Roles of Intuition. In *Intuitions*, ed. A. Booth and D. Rowbottom, 9–35. Oxford: Oxford University Press.
———. 2015. *Cognitive Phenomenology*. New York: Routledge.
———. 2017. The reality of the intuitive. *Inquiry* 60 (4): 371–385.
Clark, A. 1997. *Being There: Putting Brain, Body, and World Together Again*. Cambridge, MA: MIT Press.
Clatterbuck, H. 2013. The Epistemology of Thought Experiments: A Non-Eliminativist, Non-Platonic Account. *European Journal for Philosophy of Science* 3 (3): 309–329.
Cling, A.D. 2014. The Epistemic Regress Problem, the Problem of the Criterion, and the Value of Reasons. *Metaphilosophy* 45 (2): 161–171.
Cohen, L.J. 1986. *The Dialogue of Reason: An Analysis of Analytical Philosophy*. Oxford: Clarendon Press.
Cohen, S. 1988. How to be a Fallibilist. *Philosophical Perspectives* 2: 91–123.

Cohnitz, D., and S. Häggqvist. 2009. The Role of Intuitions in Philosophy. *Studia Philosophica Estonica* 2 (2): 1–14.
———. 2018. Thought Experiments in Current Metaphilosophical Debates. In *The Routledge Companion to Thought Experiments*, ed. M.T. Stuart, Y.J.H. Fehige, and J.R. Brown, 406–424. New York: Routledge.
Colaço, D., W. Buckwalter, S. Stich, and E. Machery. 2014. Eepistemic Intuitions in Fake-Barn Thought Experiments. *Episteme* 11 (2): 199–212.
Cooper, R. 2005. Thought Experiments. *Metaphilosophy* 36 (3): 328–347.
Corcilius, K. 2018. Aristotle Thought Experiments. In *The Routledge Companion to Thought Experiments*, ed. M.T. Stuart, Y.J.H. Fehige, and J.R. Brown, 57–76. London: Routledge.
Cornford, F.M. 1935. *Plato's Theory of Knowledge*. London: Kegan Paul, Trench, Trübner & Co., Ltd.
Crane, T. 2001. *Elements of Mind*. Oxford: Oxford University Press.
———. 2009. Is Perception a Propositional Attitude? *Philosophical Quarterly* 59 (236): 452–469.
Cummins, R. 1998. Reflection on Reflective Equilibrium. In *Rethinking Intuition*, ed. M. DePaul and W. Ramsey, 113–128. Lanham: Rowman & Littlefield.
Davies, D. 2018. Art and thought experiments. In *The Routledge Companion to Thought Experiments*, ed. M.T. Stuart, Y.J.H. Fehige, and J.R. Brown, 512–525. London: Routledge.
De Cruz, H. 2015. Where Philosophical Intuitions Come From. *Australasian Journal of Philosophy* 93 (2): 233–249.
Decker, J. 2012. Disagreement, Evidence, and Agnosticism. *Synthese* 187: 753–783.
Dennett, D. 1984. *Elbow room: The varieties of free will worth wanting*. Cambridge, MA: MIT Press.
———. 2013. *Intuition Pumps and Other Tools for Thinking*. New York: W. W. Norton & Company.
DePaul, M.R. 1986. Reflective equilibrium and foundationalism. *American Philosophical Quarterly* 23 (1): 59–69.
Descartes, R. 1996. *Meditations on First Philosophy: With Selections from the Objections and Replies*. Revised ed. Cambridge: Cambridge University Press.
Deutsch, M. 2009. Experimental Philosophy and the Theory of Reference. *Mind and Language* 24 (4): 445–466.
———. 2010. Intuitions, Counter-Examples, and Experimental Philosophy. *Review of Philosophy and Psychology* 1 (3): 447–460.
———. 2015. *The Myth of the Intuitive: Experimental Philosophy and Philosophical Method*. Cambridge, MA: MIT Press.
———. 2016. Arguments, Intuitions, and Philosophical Cases: A Note on A Metaphilosophical Dialectic. *Philosophical Forum* 47 (3-4): 297–307.
Devitt, M. 2006. Intuitions in Linguistics. *British Journal for the Philosophy of Science* 57: 481–513.
Dogramaci, S., and S. Horowitz. 2016. An Argument for Uniqueness About Evidential Support. *Philosophical Issues* 26 (1): 130–147.
Dohrn, D. 2018. Thought experiments without possible worlds. *Philosophical Studies* 175: 363–384.
Douven, I. 2009. Uniqueness Revisited. *American Philosophical Quarterly* 46: 347–361.
Dreyfus, G. 1997. *Recognizing Reality: Dharmakirti's Philosophy and Its Tibetan Interpretations*. New York: SUNY Press.
Earlenbaugh, J., and B. Molyneux. 2009. Intuitions Are Inclinations to Believe. *Philosophical Studies* 145: 89–109.
———. 2010. If intuitions must be evidential then philosophy is in big trouble. *Studia Philosophica Estonica* 2 (2): 35–53.
Elswyk, P.v. 2019. Propositional anaphors. *Philosophical Studies* 176: 1055–1075.

Engel, P. 2011. Philosophical thought experiments: in or out of the armchair? In *Thought Experiments in Methodological and Historical Contexts*, ed. K. Ierodiakonou and S. Roux, 145–163. Leiden/Boston: Brill.
Ennis, R.H. 1984. Problems in Testing Informal Logic Critical Thinking Reasoning Ability. *Informal Logic* 9 (1): 3–9.
Fedyk, M. 2009. Philosophical Intuitions. *Studia Philosophica Estonica* 2 (2): 54–80.
Fehige, Y. 2018. Theology and Thought Experiments. In *The Routledge Companion to Thought Experiments*, ed. M.T. Stuart, Y.J.H. Fehige, and J.R. Brown, 183–194. London: Routledge.
Feldman, R. 1994. Good Arguments. In *Socializing Epistemology: The Social Dimensions of Knowledge*, ed. F. Schmitt, 159–199. Lanham, MD: Rowman & Littlefield.
———. 2006. Epistemological Puzzles about Disagreement. In *Epistemology Futures*, ed. S. Hetherington, 216–236. Oxford: Oxford University Press.
———. 2007. Reasonable Religious Disagreements. In *Philosophers Without Gods: Meditations on Atheism and the Secular Life*, ed. L. Antony, 194–214. Oxford: Oxford University Press.
———. 2011. Reasonable Religious Disagreement. In *Social Epistemology: Essential Reading*, ed. A.I. Goldman and D. Whitcomb, 137–157. New York: Oxford University Press.
Feltz, A. 2008. *The Relevance of Folk Intuitions to Philosophical Debates*, Ph.D. dissertation. Tallahassee: Florida State University.
Fodor, J.A. 1975. *The Language of Thought*. New York: Crowell.
———. 1978. Propositional Attitudes. *The Monist* 61 (4): 501–523.
Foot, P. 2002. The Problem of Abortion and the Doctrine of the Double Effect. In *Virtues and Vices and Other Essays in Moral Philosophy*, 19–32. Oxford: Oxford University Press.
Forbes, G. 2000. Objectual Attitudes. *Linguistics and Philosophy* 23: 141–183.
Forge, J. 1991. Thought Experiments in the Philosophy of Physical Science. In *Thought Experiments in Science and Philosophy*, ed. T. Horowitz and G. Massey, 209–222. Lanham: Rowman and Littlefield.
Frege, G. 1956. The Thought: A Logical Inquiry. *Mind* 65: 289–311.
Gendler, T.S. 2007. Philosophical thought experiments, intuitions, and cognitive equilibrium. *Midwest Studies in Philosophy* 31 (1): 68–89.
Gettier, E.I. 1963. Is Justified True Belief Knowledge? *Analysis* 23: 121–123.
Ghins, M. 2007. Laws of Nature: Do We Need a Metaphysics? *Principia* 11: 127–150.
Glenberg, A.M. 1997. What Memory Is for. *Behavioral and Brain Sciences* 20 (1): 1–19.
Glock, H.-J. 2017. Impure Conceptual Analysis. In *The Cambridge Companion to Philosophical Methodology*, ed. G. D'Oro and S. Overgaard, 77–100. Cambridge: Cambridge University Press.
Goffi, J.-Y., and S. Roux. 2018. A Dialectical Account of Thought Experiments. In *The Routledge Companion to Thought Experiments*, ed. M.T. Stuart, Y.J.H. Fehige, and J.R. Brown, 439–453. London: Routledge.
Goldman, A.I. 1976. Discrimination and Perceptual Knowledge. *The Journal of Philosophy* 73: 771–791.
———. 1993. The Psychology of Folk Psychology. *Behavioral and Brain Sciences* 16 (1): 15–28.
———. 1994. Argumentation and Social Epistemology. *The Journal of Philosophy* 91 (1): 27–49.
———. 1997. Argumentation and Interpersonal Justification. *Argumentation* 11 (2): 155–164.
———. 2003. An Epistemological Approach to Argumentation. *Informal Logic* 23 (1): 51–63.
———. 2007. Philosophical Intuitions: Their Target, Their Source, and Their Epistemic Status. *Grazer Philosophische Studien* 74: 1–26.
———. 2010. Philosophical naturalism and intuitional methodology. *American Philosophical Association* 84 (2): 117–141.
———. 2013. Philosophical Naturalism and Intuitional Methodology. In *The a Priori in Philosophy*, ed. A. Casullo and J.C. Thurow, 11–44. Oxford: Oxford University Press.
Goldman, A.I., and J. Pust. 1998. Philosophical Theory and Intuitional Evidence. In *Rethinking Intuition: The Psychology of Intuition and its Role in Philosophical Inquiry*, ed. M. Depaul and W. Ramsey, 179–197. Lanham: Rowman & Littlefield.

Gopnik, A., and E. Schwitzgebel. 1998. Whose Concepts Are They, Anyway? The Role of Philosophical Intuition in Empirical Psychology. In *Rethinking Intuition: The Psychology of Intuition and Its Role in Philosophical Inquiry*, ed. M. DePaul and W. Ramsey, 75–91. Lanham: Rowman and Littlefield.

Govier, T. 1992. What Is a Good Argument? *Metaphilosophy* 23: 393–409.

———. 2010. *A Practical Study of Argument*. 7th ed. Belmont, CA: Wadsworth.

Grundmann, T. 2018. Platonism and the a Priori in Thought Experiments. In *The Routledge Companion to Thought Experiments*, ed. M.T. Stuart, Y.J.H. Fehige, and J.R. Brown, 293–308. London: Routledge.

Grzankowski, A. 2012. Not All Attitudes Are Propositional. *The European Journal of Philosophy* 23 (3): 374–391.

———. 2013. Non-Propositional Attitudes. *Philosophy Compass* 8 (12): 1123–1137.

———. 2016a. Limits of Propositionalism. *Inquiry* 59 (7-8): 819–838.

———. 2016b. Attitudes Towards Objects. *Noûs* 50 (2): 314–328.

Häggqvist, S. 1996. *Thought Experiments in Philosophy*. Stockholm: Almqvist and Wiksell International.

———. 2019. Thought Experiments, Formalization, and Disagreement. *Topoi* 38: 801–810.

Hales, S.D. 2000. The Problem of Intuition. *American Philosophical Quarterly* 37 (2): 135–147.

———. 2006. *Relativism and the Foundations of Philosophy*. Cambridge: MIT Press.

———. 2012. The Faculty of Intuition. *Analytic Philosophy* 53 (2): 180–207.

Hamdo, M. 2018. Epistemology Without Intuition. *International Journal of Innovative Studies in Sociology and Humanities* 3 (10): 49–53.

Hannon, M. 2018. Intuitions, Reflective Judgments, and Experimental Philosophy. *Synthese* 195: 4147–4168.

Henderson, D., and T. Horgan. 2013. On the Armchair Justification of Conceptually Grounded Necessary Truths. In *The a Priori in Philosophy*, ed. A. Casullo and J.C. Thurow, 111–133. Oxford: Oxford University Press.

Hill, C.S. 2013. Concepts, Teleology, and Rational Revision. In *The a Priori in Philosophy*, ed. A. Casullo and J.C. Thurow, 134–157. Oxford: Oxford University Press.

Hintikka, J., and M.B. Hintikka. 1989. *The Logic of Epistemology and the Epistemology of Logic: Selected Essays*. Dordrecht: Kluwer Academic Publishers.

Hoffmann, M.H. 2005. Limits of Truth: Exploring Epistemological Approaches to Argumentation. *Informal Logic* 25 (3): 245–260.

Hogarth, R.M. 2001. *Educating Intuition*. Chicago: University of Chicago Press.

Holley, D.M. 2013. Religious Disagreements and Epistemic Rationality. *International Journal for Philosophy of Religion* 74 (1): 33–48.

Horowitz, S. 2014. Immoderately Rational. *Philosophical Studies* 167: 41–56.

Huemer, M. 2001. *Skepticism and the Veil of Perception*. Lanham: Rowman and Littlefield.

———. 2005a. *Moral Intuitionism*. New York: Palgrave Macmillan.

———. 2005b. *Ethical Intuitionism*. New York: Palgrave Macmillan.

Husserl, E. 1970. *Logical Investigations*. New York: Humanities Press.

———. 1983. *Ideas Pertaining to a Pure Phenomenology and to a Phenomenological Philosophy: First Book: General Introduction to a Pure Phenomenology*. 1st ed. Dordrecht: Springer.

———. 2001. *Logical Investigations*. London: Routledge.

Ichikawa, J.J. 2013. Experimental Philosophy and Apriority. In *The a Priori in Philosophy*, ed. A. Casullo and J.C. Thurow, 45–66. Oxford: Oxford University Press.

Ichikawa, J., and B. Jarvis. 2009. Thought-Experiment Intuitions and Truth in Fiction. *Philosophical Studies* 142: 221–246.

Ichikawa, J.J., and B.W. Jarvis. 2013. *The rules of thought*. Oxford: Oxford University Press.

Ierodiakonou, K. 2011. Remarks on the History of an Ancient Thought Experiment. In *Thought Experiments in Methodological and Historical Contexts*, ed. K. Ierodiakonou and S. Roux, 37–49. Leiden/Boston: Brill.

———. 2018. The triple life of thought experiments. In *The Routledge Companion to Thought Experiments*, ed. M.T. Stuart, Y.J.H. Fehige, and J.R. Brown, 31–43. London: Routledge.

Ikeya, A. 1995. Predicate-Argument Structure of English Adjectives. *Information and Computation, Volume Proceedings of the 10th Pacific Asia Conference on Language*, pp. 149–156.

Inwagen, P.v. 1997. Materialism and the Psychological-Continuity Account of Personal Identity. *Philosophical Perspectives* 11: 305–319.

Irvine, A. 1991. Thought Experiments in Scientific Reasoning. In *Thought Experiments in Science and Philosophy*, ed. T. Horowitz and G. Massey, 149–165. Lanham: Rowman and Littlefield.

Jackson, F. 1982. Epiphenomenal Qualia. *The Philosophical Quarterly* 32: 127–136.

———. 1986. What Mary Didn't Know. *The Journal of Philosophy* 83: 291–295.

James, W. 2000. *Pragmatism and Other Writings*. London: Penguin Books.

Järvilehto, L. 2011. *Pragmatic a Priori Knowledge: A Pragmatic Approach to the Nature and Object of What Can Be Known Independently of Experience*. s.l.: Jyväskylä: University of Jyväskylä, Ph.D. thesis.

Jenkin, Z. 2020. The Epistemic Role of Core Cognition. *The Philosophical Review* 129 (2): 251–298.

Jenkins, C.S. 2007. Epistemic Norms and Natural Facts. *American Philosophical Quarterly* 44: 259–272.

———. 2008. *Grounding Concepts: An Empirical Basis for Arithmetical Knowledge*. Oxford: Oxford University Press.

———. 2013. Naturalistic Challenges to the A Priori. In *The a Priori in Philosophy*, ed. A. Casullo and J.C. Thurow, 274–290. Oxford: Oxford University Press.

Juthe, A. 2005. Argument by Analogy. *Argumentation. Argumentation* 19: 1–27.

Kahneman, D. 2011. *Thinking, Fast And Slow*. New York: Farrar, Straus And Giroux.

Kant, I. 1998. *Critique of Pure Reason*. Cambridge: Cambridge University Press.

Kelly, T. 2010. Peer Disagreement and Higher Order Evidence. In *Disagreement*, ed. R. Feldman and T.A. Warfield, 111–174. Oxford: Oxford University Press.

———. 2013. Evidence Can Be Permissive. In *Contemporary Debates in Epistemology*, ed. M. Steup, J. Turri, and E. Sosa, 2nd ed., 298–311. Hoboken: Wiley.

Kelly, T., and S. McGrath. 2010. Is reflective equilibrium enough? *Philosophical Perspectives* 24 (1): 325–359.

Khalidi, M.A. 2009. How Scientific Is Scientific Essentialism? *Journal for General Philosophy of Science* 40: 85–101.

Kidd, C. 2014. Husserl's Phenomenological Theory of Intuition. In *Rational Intuition: Philosophical Roots, Scientific Investigations*, ed. L. Osbeck and B. Held, 131–150. New York: Cambridge University Press.

Kim, M., and Y. Yuan. 2015. No Cross-Cultural Differences in Gettier Car Case Intuition: A Replication Study of Weinberg et al. 2001. *Episteme* 12 (3): 355–361.

Knobe, J. 2010. Person as scientist, person as moralist. *Behavioral and Brain Sciences* 33 (4): 315–329.

Koksvik, O. 2011. *Intuition*. Ph.D. thesis. Canberra: Australian National University.

Koksvik, O. 2017. The Phenomenology of Intuition. *Philosophy Compass* 12 (1): 1–11.

Kopec, M. 2015. A Counterexample to the Uniqueness Thesis. *Philosophia* 43 (2): 403–409.

Kopec, M., and M.G. Titelbaum. 2016. The Uniqueness Thesis. *Philosophy Compass* 11 (4): 189–200.

Kornblith, H. 1998. The Role of Intuition in Philosophical Inquiry: An Account with No Unnatural Ingredients. In *Rethinking Intuition: The Psychology of Intuition and its Role in Philosophical Inquiry*, ed. M. DePaul and W. Ramsey, 129–142. Lanham: Rowman and Littlefield.

———. 2006. Appeals to Intuition and the Ambitions of Epistemology. In *Epistemology Futures*, ed. S. Hetherington, 10–25. Oxford: Oxford University Press.

Kratzer, A. 2006. *Decomposing attitude verbs*. Jerusalem: Talk given at The Hebrew University of Jerusalem.

Kripke, S. 1980. *Naming and Necessity*. Oxford: Basil Blackwell.

Kuntz, J.R., and J.R.C. Kuntz. 2011. Surveying Philosophers About Philosophical Intuition. *Review of Philosophy and Psychology* 2 (4): 643–665.

Lakoff, G., and M. Johnson. 1980. *Metaphors We Live By*. Chicago/London: University of Chicago Press.

Lee, I.S. 1989. *A Phenomenological Study of an Experience of Intuition and Its Meaning for the Individual*. Master thesis. Vancouver: The University of British Columbia.

Lehrer, K. 1990. *Theory of Knowledge*. Colorado: Westview Press.

Lemos, N. 1994. *Intrinsic Value*. New York: Cambridge University Press.

Lenhard, J. 2018. Thought Experiments and Simulation Experiments: Exploring Hypothetical Worlds. In *The Routledge Companion to Thought Experiments*, ed. M.T. Stuart, Y.J.H. Fehige, and J.R. Brown, 484–497. London: Routledge.

Lewis, D. 1978. Truth in Fiction. *American Philosophical Quarterly* 15: 37–46.

———. 1983. *Philosophical Papers: Volume I*. New York: Oxford University Press.

Liao, S.M. 2008. A Defense of Intuitions. *Philosophical Studies* 140 (2): 247–262.

Ludwig, K. 2007. The Epistemology of Thought Experiments: First Person versus Third Person Approaches. *Midwest Studies in Philosophy* 31 (1): 128–159.

———. 2010. Intuitions and Relativity. *Philosophical Psychology* 23 (4): 427–445.

———. 2018. Thought experiments and experimental philosophy. In *The Routledge Companion to Thought Experiments*, ed. M.T. Stuart, Y.J.H. Fehige, and J.R. Brown, 385–405. London: Routledge.

Luisa, M., and D. Chiara. 1992. Possible Worlds, Counterfactuals, and Epistemic Operators. In *Knowledge, Belief, and Strategic Interaction*, ed. C. Bicchieri, D. Chiara, and M. Luisa, 155–166. s.l: Cambridge University Press.

Lumer, C. 1991. Structure and Function of Argumentations: An Epistemological Approach to Determining Criteria for the Validity and Adequacy of Argumentations. In *Proceedings of the Second International Conference on Argumentation*, ed. F.H. van Eemeren, R. Grootendorst, J.A. Blair, and C.A. Willard, 98–107. Amsterdam: SICSAT.

Machery, E. 2011. Thought Experiments and Philosophical Knowledge. *Metaphilosophy* 42 (3): 191–214.

Machery, E., R. Mallon, S. Nichols, and S. Stich. 2004. Semantics, Cross-Cultural Style. *Cognition* 92 (3): 1–12.

Malmgren, A.-S. 2006. Is There A Priori Knowledge by Testimony? *The Philosophical Review* 115 (2): 119–241.

———. 2013. A Priori Testimony Revisited. In *The a Priori in Philosophy*, ed. A. Casullo and J.C. Thurow, 158–185. Oxford: Oxford University Press.

Mandler, J.M. 1992. How to Build a Baby: II. Conceptual primitives. *Psychological Review* 99 (4): 587–604.

Martin, A. 2001. Functional Neuroimaging of Semantic Memory. In *Handbook of Functional Neuroimaging of Cognition*, ed. R. Cabeza and A. Kingstone, 153–186. Cambridge: MIT Press.

Martin, A., and L.L. Chao. 2001. Semantic Memory and the Brain: Structure and processes. *Current Opinion in Neurobiology* 11 (2): 194–201.

Martin, A., L.G. Ungerleider, and J.V. Haxby. 2000. Category-Specificity and the Brain: The Sensory-Motor Model of Semantic Representations of Objects. In *The New Cognitive Neurosciences*, ed. M.S. Gazzaniga, 2nd ed., 1023–1036. Cambridge: MIT Press.

Matheson, J. 2011. The Case for Rational Uniqueness. *Logos & Episteme* 2 (3): 359–373.

———. 2015. *The Epistemic Significance of Disagreement*. New York: Palgrave Macmillan.

Matilal, B.K. 1986. *Perception: An Essay on Classical Indian Theories of Knowledge*. Oxford: Clarendon Press.

McAllister, J.W. 2018. Historicism and cross-culture comparison. In *The Routledge Companion to Thought Experiments*, ed. M.T. Stuart, Y.J.H. Fehige, and J.R. Brown, 425–438. London: Routledge.

McGinnis, J. 2018. Experimental thoughts about thought experiments in medieval Islam. In *The Routledge Companion to Thought Experiments*, ed. M.T. Stuart, Y.J.H. Fehige, and J.R. Brown, 77–91. London: Routledge.

McGrath, M., and Frank, D. 2018. *Propositions. s.l.:The Stanford Encyclopedia of Philosophy.*

Meacham, C. 2014. Impermissive Bayesianism. *Erkenntnis* 79: 1185–1217.

Meehan, W. 2014. A Grain of Sand: Spinoza's Conception of Intuition. In *Rational Intuition: Philosophical Roots, Scientific Investigations*, ed. L.M. Osbeck and B.S. Held, 90–113. New York: Cambridge University Press.

Merricks, T. 2009. Propositional Attitudes? *Proceedings of the Aristotelian Society* 109: 207–232.

Meynell, L. 2014. Imagination and Insight: A New Acount of the Content of Thought Experiments. *Synthese* 191: 4149–4168.

———. 2018. Images and Imagination in Thought Experiments. In *The Routledge Companion to Thought Experiments*, ed. M.T. Stuart, Y.J.H. Fehige, and J.R. Brown, 498–511. London: Routledge.

Miller, R.B. 2000. Without Intuitions. *Metaphilosophy* 31 (3): 231–250.

Miščević, N. 2018. Thought Experiments in Political Philosophy. In *The Routledge Companion to Thought Experiments*, ed. M.T. Stuart, Y.J.H. Fehige, and J.R. Brown, 153–170. London: Routledge.

Mizrahi, M. 2016. Why Gettier Cases Are Misleading. *Logos and Episteme* 7 (1): 31–44.

Moltmann, F. 2013. *Abstract Objects and Semantics of Natural Language*. Oxford: Oxford University Press.

Molyneux, B. 2014. New Arguments That Philosophers Don't Treat Intuitions as Evidence. *Metaphilosophy* 45 (3): 441–461.

Montague, M. 2007. Against Propositionalism. *Noûs* 41 (3): 503–518.

Moulton, K. 2009. *Natural Selection and the Syntax of Clausal Complementation*, Ph.D. thesis. Amherst: University of Masschusetts.

Moulton, K. 2015. CPs: Copies and Compositionality. *Linguistic Inquiry* 46 (2): 305–342.

Myers, C.M. 1986. Analytical Thought Experiments. *Metaphilosophy* 17: 109–118.

Nado, J. 2015. Philosophical Expertise and Scientific Expertise. *Philosophical Psychology* 28 (7): 1026–1044.

———. 2016. The intuition deniers. *Philosophical Studies* 173 (3): 781–800.

Nagel, T. 1979. *Mortal Questions*. Cambridge: Cambridge University Press.

Nealon, J.T. 2014. Historical A Priori. In *The Cambridge Foucault Lexicon*, ed. L. Lawlor and J. Nale, 200–206. New York: Cambridge University Press.

Nersessian, N.J. 2018. Cognitive Science, Mental Modeling, and Thought Experiments. In *The Routledge Companion to Thought Experiments*, ed. M.T. Stuart, Y.J.H. Fehige, and J.R. Brown, 309–326. London: Routledge.

Newton, N. 1996. *Foundations of Understanding*. Amsterdam: John Benjamins Publishing Company.

Nichols, S. 2004. Folk Concepts and Intuitions: From Philosophy to Cognitive Science. *Trends in Cognitive Sciences* 8 (11): 514–518.

Nichols, S., and S. Stich. 2003. How to Read Your Own Mind: A Cognitive Theory of Self-Consciousness. In *Consciousness: New Philosophical Essays*, ed. Q. Smith and A. Jokic, 157–200. Oxford: Oxford University Press.

Nichols, S., and J. Ulatowski. 2007. Intuitions and Individual Differences: The Knobe Effect Revisited. *Mind and Language* 22 (1): 346–365.

Nichols, S., S. Stich, and J.M. Weinberg. 2003. Metaskepticism: Meditations in Ethnoepistemology. In *The Skeptics*, ed. S. Luper, 227–247. Aldershot: Ashgate.

Norton, J.D. 1991. Thought Experiments in Einstein's Work. In *Thought Experiments in Science and Philosophy*, ed. G. Massey and T. Horowitz, 129–148. Lanham: Rowman & Littlefield.

———. 1993. Einstein and Nordstrom: Some Lesser-Known Thought Experiments in Gravitation. In *The Attraction of Gravitation: New Studies in the History of General Relativity*, ed. M. Janssen, J.D. Norton, and J. Earman, 3–28. Boston: Birkhauser.

———. 1996. Are Thought Experiments Just What You Thought? *Canadian Journal of Philosophy* 26: 333–366.
———. 2018. The Worst Thought Experiment. In *The Routledge Companion to Thought Experiments*, ed. M.T. Stuart, Y.J.H. Fehige, and J.R. Brown, 454–468. London: Routledge.
Oktar, S. 2014. Consciousness of the Cosmos: A Thought Experiment Through Philosophy and Science Fiction. In *Phenomenology of Space and Time*, ed. A. Tymieniecka, 149–163. Cham: Springer.
Orenstein, A. 2003. Fiction, Propositional Attitudes, and Some Truths about Falsehood. *Dialectica* 57 (Special Issue: Do Ficta Follow Fiction?): 177–190.
Palmieri, P. 2018. Galileo's Thought Experiments: Projective Participation and the Integration of Paradoxes. In *The Routledge Companion to Thought Experiments*, ed. M.T. Stuart, Y.J.H. Fehige, and J.R. Brown, 92–110. London: Routledge.
Parsons, C. 1995. Platonism and Mathematical Intuition in Kurt Gödel's Thought. *Bulletin of Symbolic Logic* 1: 44–74.
Peacock, K.A. 2018. Happiest Thought Experiments: Great Thought Experiments of Modern Physics. In *The Routledge Companion to Thought Experiments*, ed. M.T. Stuart, Y.J.H. Fehige, and J.R. Brown, 211–242. London: Routledge.
Peels, R., and A. Booth. 2014. Why Responsible Belief Is Permissible Belief. *Analytic Philosophy* 55 (1): 75–88.
Perry, J. 1994. Intentionality (2). In *A Companion to the Philosophy of Mind*, ed. S.D. Guttenplan, 386–395. Oxford: Blackwell.
Piętka, D. 2015. The Concept of Intuition and Its Role in Plato and Aristotle. *Organon* 47: 23–40.
Pinto, R.C. 2001. *Argument, Inference and Dialectic*. Dordrecht: Kluwer.
Pitt, D. 2004. The Phenomenology of Cognition or What Is It Like to Think That P? *Philosophy and Phenomenological Research* 69 (1): 1–36.
Plantinga, A. 1993. *Warrant and Proper Function*. Oxford: Oxford University Press.
Pollock, J.L. 1974. *Knowledge and Justification*. Princeton/London: Princeton University Press.
Pust, J. 2000. *Intuitions as Evidence*. New York: Garland/Routledge.
———. 2013. Skepticism, Reason, and Reidianism. In *The a Priori in Philosophy*, ed. A. Casullo and T.C. Thurow, 205–225. Oxford: Oxford University Press.
Putnam, H. 1973. Meaning and Reference. *The Journal of Philosophy* 70: 699–711.
———. 1975. The Meaning of 'Meaning. *Minnesota Studies in the Philosophy of Science* 7: 131–193.
Pylyshyn, Z.W. 1984. *Computation and Cognition: Toward a Foundation for Cognitive Science*. Cambridge, MA: Bradford Books, MIT Press.
Reiss, J. 2018. Thought Experiments and Idealizations. In *The Routledge Companion to Thought Experiments*, ed. M.T. Stuart, Y.J.H. Fehige, and J.R. Brown, 469–483. London: Routledge.
Rescher, N. 1991. Thought Experimentation in Presocratic Philosophy. In *Thought Experiments in Science and Philosophy*, ed. T. Horowitz and G.J. Massey, 31–41. Lanham, MD: Rowman & Littlefield.
Richard, M. 1990. *Propositional Attitudes: An Essay on Thoughts and How We Ascribe Them*. Cambridge: Cambridge University Press.
Robbins, P., and A.I. Jack. 2006. The Phenomenal Stance. *Philosophical Studies* 127: 59–85.
Rorty, R. 1979. *Philosophy and the Mirror of Nature*. Princeton: Princeton University Press.
———. 1984. The Historiography of Philosophy: Four Genres. In *Philosophy in History: Essays in the Historiography of Philosophy*, ed. R. Rorty, J.B. Schneewind, and Q. Skinner, 49–76. Cambridge: Cambridge University Press.
Rosa, L. 2012. Justification and the Uniqueness Thesis. *Logos & Episteme* 3 (4): 571–577.
Rosen, G. 2001. Nominalism. *Naturalism, Epistemic Relativism. Philosophical Perspectives* 15: 69–91.
Russell, B. 1919. The Philosophy of Logical Atomism. *The Monist* 29: 32–63.
Sankey, H. 2014. Relativism, Particularism and Reflective Equilibrium. *Journal for General Philosophy of Science* 45 (2): 281–292.

Schabas, M. 2018. Thought Experiments in Economics. In *The Routledge Companion to Thought Experiments*, ed. M.T. Stuart, Y.J.H. Fehige, and J.R. Brown, 171–182. London: Routledge.

Schechter, J. 2010. The Reliability Challenge and the Epistemology of Logic. *Philosophical Perspectives* 24 (1): 437–464.

Schick, T., and L. Vaughn. 2012. *Doing Philosophy: An Introduction Through Thought Experiments*. 5th ed. New York: McGraw-Hill.

Schlaepfer, G., and M. Weber. 2018. Thought Experiments in Biology. In *The Routledge Companion to Thought Experiments*, ed. M.T. Stuart, Y.J.H. Fehige, and J.R. Brown, 243–256. London: Routledge.

Schoenfield, M. 2014. Permission to Believe: Why Permissivism Is True and What It Tells Us About Irrelevant Influences on Belief. *Noûs* 48 (2): 193–218.

Searle, J.R. 1980. Minds, Brains, and Programs. *Behavioral and Brain Sciences* 3 (3): 417–457.

Seyedsayamdost, H. 2015. On Gender and Philosophical Intuition: Failure of Replication and Other Negative Results. *Philosophical Psychology* 28 (5): 642–673.

Sinnott-Armstrong, W., and R.J. Fogelin. 2013. *Understanding Arguments: An Introduction to Informal Logic*. 9th ed. Stamford: Cengage Learning.

Smith, E.E., and D.L. Medin. 1981. *Categories and Concepts*. Cambridge: Harvard University Press.

Sorensen, R.A. 1992. *Thought Experiments*. New York: Oxford University Press.

Sosa, E. 1996. Rational Intuition: Bealer on Its Nature and Epistemic Status. *Philosophical Studies* 81: 151–162.

———. 1998. Minimal Intuition. In *Rethinking Intuition: The Psychology of Intuition and Its Role in Philosophical Inquiry*, ed. M. DePaul and W. Ramsey, 257–270. Lanham: Rowman and Littlefield.

———. 2003. Chisholm's Epistemic Principles. *Metaphilosophy* 34: 553–562.

———. 2006. Intuitions and Truth. In *Truth and Realism*, ed. P. Greenough and M.P. Lynch, 208–226. Oxford: Oxford University Press.

———. 2007a. *A Virtue Epistemology: Apt Belief and Reflective Knowledge, Volume I*. Oxford/ New York: Oxford University Press.

———. 2007b. Intuitions: Their Nature and Epistemic Efficacy. *Grazer Philosophische Studien* 74: 51–67.

———. 2007c. Experimental Philosophy and Philosophical Intuition. *Philosophical Studies* 132: 99–107.

———. 2009. A Defense of the Use of Intuitions in Philosophy. In *Stich and His Critics*, ed. M. Bishop and D. Murphy, 101–112. Chichester: Blackwell.

———. 2010. Intuitions and Meaning Divergence. *Philosophical Psychology* 23 (4): 419–426.

———. 2011. Can There Be a Discipline of Philosophy? And Can It Be Founded on Intuitions? *Mind & Language* 26: 453–467.

———. 2013. Intuitions and Foundations: The Relevance of Moore and Wittgenstein. In *The a Priori in Philosophy*, ed. A. Casullo and J.C. Thurow, 186–202. Oxford: Oxford University Press.

———. 2014. Intuitions, Their Nature and Probative Value. In *Intuitions*, ed. A.R. Booth and D.P. Rowbottom, 36–50. Oxford: Oxford University Press.

Spicer, F. 2010. Cultural Variations in Folk Epistemic Intuitions. *Review of Philosophy and Psychology* 1 (4): 515–529.

Starikova, I., and M. Giaquinto. 2018. Thought Experiments in Mathematics. In *The Routledge Companion to Thought Experiments*, ed. M.T. Stuart, Y.J.H. Fehige, and J.R. Brown, 257–278. London: Routledge.

Stich, S., and K. Tobia. n.d. Intuition and its critics. In *The Routledge Companion to Thought Experiments*, ed. M.T. Stuart, Y.J.H. Fehige, and J.R. Brown, 369–384. London: Routledge.

Stoljar, D. 1996. Nominalism and Intentionality. *Noûs* 30: 221–241.

Stoltz, J. 2007. Gettier and Factivity in Indo-Tibetan Epistemology. *The Philosophical Quarterly* 57: 394–415.

Strawson, P.F. 1952. *Introduction to Logical Theory*. London: Methuen.

Stuart, M.T. 2016. Norton and the Logic of Thought Experiments. *Axiomathes* 26: 451–466.

———. 2018. How Thought Experiments Increase Understanding. In *The Routledge Companion to Thought Experiments*, ed. M.T. Stuart, Y.J.H. Fehige, and J.R. Brown, 526–544. London: Routledge.

Swain, S., J. Alexander, and J.M. Weinberg. 2008. The Instability of Philosophical Intuitions: Running Hot and Cold on Truetemp. *Philosophy and Phenomenological Research* 76 (1): 138–155.

Talbot, B. 2009. Psychology and the Use of Intuitions in Philosophy. *Studia Philosophica Estonica* 2 (2): 157–176.

Taliaferro, C., and E. Knuths. 2017. Thought Experiments in Philosophy of Religion: The Virtues of Phenomenological Realism and Values. *Open Theology* 3 (1): 167–173.

Thagard, P. 2006. Desires Are Not Propositional Attitudes. *Dialogue* 45 (1): 151–156.

Thomson, J.J. 1985. The trolley problem. *The Yale Law Journal* 94: 1395–1415.

Thurow, J.C. 2013. The Implicit Conception and Intuition Theory of the A Priori, with Implications for Experimental Philosophy. In *The a Priori in Philosophy*, ed. A. Casullo and J.C. Thurow, 67–91. Oxford: Oxford University Press.

Tidman, P. 1996. The Justification of A Priori Intuitions. *Philosophy and Phenomenological Research* 56: 161–171.

Titelbaum, M. 2010. Not Enough There There: Evidence, Reasons, and Language Independence. *Philosophical Perspectives* 24 (1): 477–528.

Tulving, E. 1972. Episodic and semantic memory. In *Organization and Memory*, ed. E. Tulving and W. Donaldson, 381–403. New York: Academic.

Vaesen, K., M. Peterson, and B.V. Bezooijen. 2013. The Reliability of Armchair Intuitions. *Metaphilosophy* 44 (5): 559–578.

Weber, M.A. 2017. Armchair Disagreement. *Metaphilosophy* 48 (4): 527–549.

Wedgwood, R. 2013. A Priori Bootstrapping. In *The a Priori in Philosophy*, ed. A. Casullo and J.C. Thurow, 226–246. Oxford: Oxford University Press.

Weinberg, J.M. 2007. How to Challenge Intuitions Empirically Without Risking Skepticism. *Midwest Studies in Philosophy* 31 (1): 318–343.

———. 2013. The Prospects for an Experimentalist Rationalism, or Why It's OK if the A Priori Is Only 99.44 Percent Empirically Pure. In *The a Priori in Philosophy*, ed. A. Casullo and J.C. Thurow, 92–108. Oxford: Oxford University Press.

Weinberg, J.M., S. Nichols, and S. Stich. 2001. Normativity and Epistemic Intuitions. *Philosophical Topics* 29: 429–460.

Weinberg, J.M., et al. 2012. Intuition & Calibration. *Essays in Philosophy* 13 (1): 257–284.

Westphal, K.R. 2018. Thought Experiments, Epistemology and Our Cognitive (in) Capacities. In *The Routledge Companion to Thought Experiments*, ed. M.T. Stuart, Y.J.H. Fehige, and J.R. Brown, 128–149. London: Routledge.

White, R. 2005. Epistemic Permissiveness. *Philosophical Perspectives* 19 (1): 445–459.

———. 2007. Epistemic Subjectivism. *Episteme* 4 (1): 115–129.

———. 2013. Evidence Cannot Be Permissive. In *Contemporary Debates in Epistemology*, ed. M. Steup, J. Turri, and E. Sosa, 2nd ed., 312–323. Hoboken: Wiley.

Williamson, T. 2000. *Knowledge and its Limits*. Oxford: Oxford University Press.

———. 2002. *Knowledge and Its Limits*. Oxford: Oxford University Press.

———. 2004. Philosophical 'Intuitions' and Scepticism about Judgement. *Dialectica* 58 (1): 109–153.

———. 2007a. *The Philosophy of Philosophy*. Hoboken: Wiley-Blackwell.

———. 2007b. Philosophical Knowledge and Knowledge of Counterfactuals. *Grazer Philosophische Studien* 74 (1): 89–123.

———. 2011. Philosophical Expertise and the Burden of Proof. *Metaphilosophy* 42 (3): 215–229.

———. 2013a. How Deep is the Distinction between A Priori and A Posteriori Knowledge? In *The a Priori in Philosophy*, ed. A. Casullo and J.C. Thurow, 291–312. Oxford: Oxford University Press.

———. 2013b. Gettier Cases in Epistemic Logic. *Inquiry* 56 (1): 1–14.

———. 2016. Philosophical Criticisms of Experimental Philosophy. In *A Companion to Experimental Philosophy*, ed. J. Sytsma and W. Buckwalter, 22–36. Oxford: Blackwell.

Wiltsche, H.A. 2018. Phenomenology and Thought Experiments: Thought experiments as anticipation pumps. In *The Routledge Companion to Thought Experiments*, ed. M.T. Stuart, Y.J.H. Fehige, and J.R. Brown, 342–365. London: Routledge.

Wittgenstein, L. 1922. *Tractatus Logico-Philosophicus*. London/New York: Trench, Trubner & CO., LTD/Harcourt, Brace, & Company, Inc.

———. 1958. *Philosophical Investigations*. 3rd ed. Oxford: Basil Blackwell Ltd.

———. 1976. *Wittgenstein's Lectures on the Foundations of Mathematics, Cambridge 1939*. Chicago/London: The University of Chicago Press.

———. 1978. *Remarks on the Foundations of Mathematics*. 3rd ed. Oxford: Blackwell.

Witt-Hansen, J. 1976. H.C. Ørsted, Immanuel Kant, and the Thought Experiment. *Danish Yearbook of Philosophy* 13: 48–65.

Wright, J.C. 2010. On Intuitional Stability: The Clear, the Strong, and the Paradigmatic. *Cognition* 115 (3): 491–503.

Index

M. Hamdo, *Epistemic Thought Experiments and Intuitions*, Philosophical Studies Series 150, https://doi.org/10.1007/978-3-031-33480-1

V

W

Z

MIX
Papier aus verantwortungsvollen Quellen
Paper from responsible sources
FSC® C105338
www.fsc.org

If you have any concerns about our products,
you can contact us on
ProductSafety@springernature.com

In case Publisher is established outside the EU,
the EU authorized representative is:
Springer Nature Customer Service Center GmbH
Europaplatz 3, 69115 Heidelberg, Germany

Printed by Libri Plureos GmbH
in Hamburg, Germany